ELECTRON SPIN
RELAXATION PHENOMENA
IN SOLIDS

MONOGRAPHS ON ELECTRON SPIN RESONANCE

Editor: H. M. Assenheim, Israel Atomic Energy Commission

Electron Spin Relaxation Phenomena in Solids

K. J. STANDLEY

M.A., D.Phil.(Oxon.), F.Inst.P., F.R.S.E.

R. A. VAUGHAN

B.Sc., Ph.D.(Nott.)
University of Dundee

ADAM HILGER LTD
LONDON

SBN 85274 092 1

Published by
ADAM HILGER LTD
60 Rochester Place, London NW1

Printed in Great Britain by Bell and Bain Ltd, Glasgow

EDITOR'S PREFACE

The study of relaxation effects by magnetic resonance methods has developed to the point where it merits and needs comprehensive review. E.S.R. spectrometry usually gives a clear picture of relaxation mechanisms and, largely in consequence, an insight into the structure of a sample and its surroundings. Professor Standley, whose specialist group at Dundee has long had a world-wide reputation, has been joined by his colleague, Dr Vaughan, in writing the present monograph with its lucid explanations of the mechanisms and their underlying physics, theoretical and experimental. I have found the book remarkably easy and pleasant to read. I know for certain that it will appeal to all whose interests touch any aspects of relaxation and that it will deservedly take its place among the classics of electron spin resonance.

H. M. ASSENHEIM

May 1969

PREFACE

Electron spin resonance studies were first made more than two decades ago, and, naturally, the different effects of spin-lattice and spin–spin relaxation times, T_1 and T_2, were soon encountered. Something was, however, already known about this subject: from the late thirties onwards measurements had been made of magnetic susceptibility and power absorption at low temperatures, resulting in numerical values for the relaxation parameters in concentrated crystals. One associates particularly such names as Casimir, Gorter and du Pré with this work, while Van Vleck is generally recognized as presenting the first sound theoretical analysis of the problem in 1939 and 1940, developing further the earlier ideas of Fierz, Heitler and Teller, Kronig and Waller. As E.S.R. techniques improved, so the study of spin-lattice relaxation developed also, first through saturation-broadening measurements and then through the methods employing pulse saturation of the absorption line followed by its monitored recovery. There are now several techniques for the measurement of T_1 and the theory has since been further developed. It seemed, therefore, an appropriate time to review theory and experiment, and this book is an attempt to present a basic account—not necessarily a complete one in places—of spin-lattice relaxation and allied phenomena in solids.

After a general introduction to the subject, the relevant theory is discussed in Chapter 2. This relates primarily to situations where we have two spin levels only; cross relaxation, involving other spin levels, is dealt with in Chapter 3 and the phonon bottleneck problem is considered in Chapter 4. Chapters 5 to 8 concentrate upon the experimental determination of T_1; we have given an outline of the theory of each method and the way it is used in practice, an assessment of its general applicability and an estimate of the reliability of experimental results to which it may lead. In Chapter 9, to give an idea of the present state of the art, we compare typical theoretical and experimental results published on ruby (Cr^{3+} in Al_2O_3) and on certain rare-earth ions, while in Appendix 2 we have collected together other T_1 results which have been published over the last twelve years. Such a list cannot be either complete or wholly accurate in such an abbreviated form, but we hope it will be useful to other workers in the field.

A book of this type cannot be produced without the help of many people. Several friends have materially assisted by their comments.

In particular, Dr A. M. Stoneham's advice on Chapter 2 was most appreciated and that chapter is much better for the changes he suggested—any errors remaining are ours! We have, of course, used illustrations from many published papers and in each case we have acknowledged the source in the caption. May we here, collectively thank all the authors and publishers who so readily gave us permission to reproduce these illustrations? In the publisher's office, Mrs E. Jacobi interpreted our rough sketches with uncanny skill and accuracy, and Mr D. Tomlinson smoothed our path with quiet efficiency. Finally we must record our thanks to Miss P. M. Mitchell who prepared the typescript with the meticulous care which characterizes all her work.

K. J. STANDLEY
R. A. VAUGHAN

Dundee
May 1969

CONTENTS

Chapter 1

Spin-Lattice Relaxation Phenomena

More than fifty years ago, Einstein showed that induced transitions between a given pair of energy states occur with equal probability in each direction. At microwave frequencies one can consider merely the effect of a radiation field stimulating transitions in each direction with equal probability, and neglect the effect of spontaneous emission, which may be several orders of magnitude less probable. One observes resonance *absorption* (for example in electron spin resonance) because the populations of the two levels are not equal. A Boltzmann distribution occurs with the population of the lower level greater than that of the upper level. Since the number of transitions which take place in unit time from a certain level is just equal to the population of that level multiplied by the probability for each transition, there is a nett absorption of energy from the radiation field. If this were the only process taking place, clearly the populations would rapidly be equalized, and resonant absorption would cease. Since the system must return eventually to the original Boltzmann distribution when the cause of the disturbance is removed, there must be a mechanism acting to restore equilibrium; such a mechanism is called a 'relaxation process'. The relaxation phenomenon is found in many branches of physics, and occurs since a physical system will tend to reach its lowest possible energy configuration, a quiescent or equilibrium condition. The strength of a relaxation process is designated by a relaxation time or its inverse, a relaxation rate, which is just a measure of the rate at which equilibrium would be regained if there were no opposing action.

The important characteristic feature of a paramagnet is the existence of permanent magnetic dipole moments, which give rise to the energy-level scheme. In the solid state, one usually regards such a paramagnetic material as comprising two weakly-interacting subsystems. In the majority of transition metal atoms in solids the magnetic moment is due mainly to the *spin* of the electrons in the atom, and for this reason we shall use the term 'spin system' to denote one of these subsystems. The other subsystem is the surrounding crystal lattice, or rather the vibrations of the crystal lattice, often referred to just as 'the lattice'. Energy communicated from the spin system to the lattice is rapidly dissipated in a heat sink of infinite thermal capacity, the bath, with which we usually consider the lattice to be in intimate contact. This weak interaction between the spin system and the lattice provides the dominant relaxation process in electron spin resonance (E.S.R.), and is the process which we shall be mainly considering. Another relaxation process is one whereby one species of spins which is resonant with the radiation field can communicate its energy to another species of spins which for some reason is not resonant. This process, in some of its many guises, will also be mentioned, but it is often so much more rapid than spin-lattice relaxation that it is usual to consider that this is the mechanism which keeps the spin system in internal equilibrium and that this spin system as a whole then relaxes to the lattice.

The first real attempt to investigate paramagnetic relaxation was made by the Leiden group led by Gorter and Casimir in the mid 1930s. They investigated the behaviour of the paramagnetic susceptibility of the sample, when subjected to radio-frequency fields, and determined the maximum rate at which the magnetic moment of the specimen could follow the changing applied field. The situation was analysed by Casimir and du Pré[1] in terms of a thermodynamic approach to the 'conduction' of energy from the spins to the lattice. Two basic assumptions were made in this approach. One was that the spin system could always be considered to be in internal equilibrium, characterized by a single temperature T_s, defined by means of the Boltzmann distribution of spins (see p. 10). The other was that the rate of heat flow between the spin system and the lattice was proportional to their instantaneous temperature difference. The interpretation of the experimental results is dependent on the nature of the physical model assumed, and other more direct methods are now available. It was, however, the only method available for nearly twenty years, and as such it has contributed much to the basic understanding of the subject. We discuss this work further in Chapter 5.

The early theoretical papers were mainly devoted to an attempt to account for the results of such 'non-resonant' experiments. The papers of Van Vleck[2,3] laid the foundations of our present understanding of the subject. Two types of interaction between the spin system and the lattice vibrations were proposed. In the first, the *direct process*, the spins exchange quanta of energy with lattice modes having the same frequency as the spin resonance frequency. The frequencies of the lattice modes are considered distributed according to a Debye spectrum When a spin relaxes, it emits a quantum of acoustic energy (or a phonon) within a narrow band of energies, into one of these modes In the *Raman process*, however, a spin interacts simultaneously with two lattice modes whose *difference* frequency is the spin resonance frequency. This is a second-order process, and is less probable than the direct process, but the number of different combinations of frequencies satisfying the necessary resonance condition is very large. The probability of Raman-type relaxation increases very rapidly with temperature, according to a T^7 or T^9 law, whereas that of the direct process varies only linearly with temperature. The nett result of these two effects is that the direct process usually dominates in the liquid helium range of temperatures, whereas the Raman process dominates at higher temperatures. The quantitative results of these theories are discussed in Chapter 2.

The phenomena of E.S.R. are now quite well understood, and are very well documented (see, for example, Assenheim[4]), but the position of relaxation effects is not nearly as satisfactory. Attempts have been made to account theoretically for the relaxation behaviour of spin systems, with varying degrees of success, but the experimental results, although great in number, have often been obtained with an insufficient degree of control over the experimental conditions to allow any detailed analysis and reliable comparison with theory to be made. In the light of our present knowledge, it may well be the case that some of the earlier apparent inconsistencies are to be attributed to a lack of reproducibility of experimental conditions. This problem will be discussed further in later sections.

§1.1

RELAXATION PROCESSES

We shall now consider in more detail the various mechanisms by which the spin system may lose the energy given to it by the radiation field before discussing the dynamics of the spin system for the dominating spin-lattice contribution.

Energy is absorbed from the radiation field by spins (X) with which it is resonant. This is shown diagrammatically in Fig. 1.1. If there is also present a different species of spin with the same resonant frequency (X′), energy is transferred between these two systems in a time T_2, the spin–spin relaxation time, which therefore serves to keep all these spins in equilibrium. Now consider other spins (Y) having a slightly different resonant frequency. These may be different spin packets in an inhomogeneously broadened line (§8.1) or hyperfine components. Energy is transferred from the resonant spins to these Y spins by diffusion mechanisms, which may involve the exchange of quanta of energy with the lattice and which usually occur at a rate slower than T_2^{-1} but faster than T_1^{-1}. There are also other processes by which energy may be transferred between these two spin systems. If the resonant frequency of the Y spins is a multiple of that of the X spins, then *harmonic cross relaxation* (Chapter 3) may occur in which, say, two X spins flip

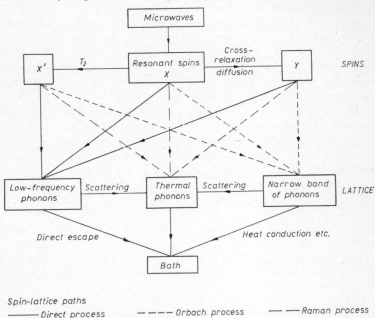

Spin-lattice paths
———— Direct process   — — — — Orbach process   — —— Raman process

FIG. 1.1. Schematic diagram of the various relaxation processes which may occur within a sample. Microwave energy supplied to the resonant spins X eventually finds its way to the surrounding bath by one or more of the indicated routes.

down and exchange energy with one Y spin flipping up. All these processes, spin-spin relaxation, spin diffusion and cross relaxation, are independent of temperature to first order.

Next consider the lattice in which the phonons are distributed according to a Debye spectrum at any particular lattice temperature. Spins X may exchange energy with the lattice by emitting a phonon at the resonant frequency v into a lattice mode of the same frequency. In this direct process only low-frequency phonons with energy hv are employed. In the Raman process, the spin interacts simultaneously with two lattice modes whose difference frequency is the spin frequency. One phonon is absorbed and another emitted, and the energy difference $hv_1 - hv_2$ is absorbed or emitted by the spin system effecting a transition. All the phonons in the Debye spectrum may now be used, subject only to this condition. In a variation of this mechanism, first envisaged by Orbach,[5] spins may be promoted to a real excited level by the absorption of a high-energy phonon. These spins then relax to the lower level in the ground doublet by emitting a phonon of slightly higher energy. This difference in energies is again lost by the spins, resulting in an effective transition between the two lower levels. This 'resonant two-phonon process' is only possible if there is another energy level at $\Delta_c < k\theta_D$ (θ_D is the Debye temperature) so that phonons are available to cause this transition. This process differs from the Raman process in two very important details. Firstly, the Raman process takes place via a *virtual* intermediate level. Secondly, *any* virtual level will suffice for the Raman process and consequently many phonons can contribute, whereas the Orbach process uses only a narrow band of phonons at $hv = \Delta_c$; now v is a high frequency, unlike in the direct process, and so there are many more of these phonons available.

The temperature variations for the various processes are:

Direct	Raman	Orbach	
$\dfrac{1}{T_1} = AT \text{ (or } A''H^2T)$	$+BT^7$	$+C \exp(-\Delta_c/kT)$	Even number of spins, non-Kramers salt.
$\dfrac{1}{T_1} = A'H^4T$	$+B'T^9 + B''H^2T^7$	$+C' \exp(-\Delta_c/kT)$	Odd number of spins, Kramers salt.

The coefficients, A, A′, etc., are such that only the first term is usually significant at helium temperatures while the Raman process is more effective at higher temperatures. All X and Y spins can relax by means of these three processes.

Now consider energy exchange within the lattice. The energy in the low-frequency phonons is only slowly exchanged with the rest of the phonons, by scattering mechanisms, etc., and so energy given to the lattice at any one frequency tends to stay at that frequency. One way in which energy may be lost by the vibrational modes is by scattering at the boundaries of the crystal, when the energy is given to the bath. This lattice-bath contact is usually assumed to be very good, so that all the energy given to the lattice by the spins is immediately transferred to the bath. This may not in fact be the case and indeed it should not be so. In experiments with acoustic phonons, a pulse of these phonons is sent into a specimen and this may be reflected up and down the specimen many times (it is not uncommon to observe a hundred echoes), showing that very little energy is lost on reflection at the crystal boundary. Also, these phonons remain in a fairly narrow frequency band, showing that not much energy is exchanged with the thermal phonons. There should therefore be a build-up in the number of phonons in those modes 'on speaking terms' with the spins, which should cause a 'spike' on the phonon spectrum. This effectively 'heats up' the phonon spectrum at that frequency (i.e. the number of phonons present at that frequency corresponds to the equilibrium number which *would* be present at a higher lattice temperature—the lattice is not itself heated up). This has the effect of slowing down the relaxation process, which is proportional to the temperature difference between the spins and the phonons, since the spins behave as though they are relaxing to a hotter lattice than that which is really present. This effect is referred to as a *phonon bottleneck*, and gives rise to a temperature dependence of the form

$$1/T_1 \propto T^2$$

The physical size of the crystal ought to be an important parameter in such a case, since this determines the time taken for a phonon to reach the surface and escape. The concentration of paramagnetic ions should also be important because the more spins there are the more severe the bottleneck should be.

This scarcity of low-frequency lattice oscillators should, in nearly all cases, be very severe, from a consideration of the number of phonons present at a frequency v and the rate at which they escape. If allowance is made for most of these phonons being reflected at the crystal boundary, then the situation is even worse, especially as the acoustic experiments indicate that specular reflection of phonons can occur without their frequency being much altered. Why then is a bottleneck not always observed? There must be some scattering mechanism present whereby spins can exchange energy with higher-frequency phonons thus by-passing the bottleneck. This mechanism is at present not well understood.

The thermal phonons also give up their energy to the bath. If there are many low-frequency phonons present at frequency v, then, as these diffuse through the lattice, they may exchange energy once again with spins at that resonant frequency and re-excite a spin. This is called phonon pumping, and tends to slow down the observed recovery time of the spins since it reduces the nett number of spins which have relaxed.

This brief survey shows the complexity of the problem, which we shall discuss in later chapters.

§1.2

DYNAMICS OF THE SPIN SYSTEM

There are two basic approaches to describing spin-lattice relaxation phenomena. One is to use the concept of transition probabilities between populated energy levels and to obtain expressions for the absorption and its time derivative, the so-called rate equations, and the other is to follow the phenomenological treatment of Bloch.[6] We shall first consider these equations which were originally formulated to describe the dynamic magnetic behaviour of nuclear spins, but apply in principle also to electron spins. Historically Bloch was the first to introduce the concept of relaxation into the macroscopic aspect of magnetic resonance.

The Bloch equations

Consider a system of magnetic moments, a spin system, which experiences simultaneously a steady field H_0 and at right angles to this a radio-frequency field H_1 at the resonant frequency $\omega_0 = \gamma H_0$, where γ is the gyromagnetic ratio. After a sufficient time, the moments will all precess in phase about the direction of H_0. If H_1 is now removed, the moments will continue to precess about H_0, but the angle of the precessional cone decreases and the magnetization tends towards its equilibrium value. This process takes place in a time characterized by T_1, the spin-lattice relaxation time.

The moments, however, may be influenced not only by the steady applied field H_0 but also by the fields due to neighbouring magnetic moments. These produce small local contributions to the steady magnetic field, and, as a consequence, each moment precesses in a slightly different field from the others. Each moment will therefore possess a different precessional frequency, and they will rapidly lose phase coherence after the removal of the radiation field. The rate at which coherence is lost is characterized by the spin-spin relaxation time T_2.

If $\Delta\omega$ is this spread in precessional frequencies, we can write

$$T_2 \sim \frac{1}{\Delta\omega} \sim \frac{1}{\gamma\Delta H_0}$$

where ΔH_0 is the spread in local fields present, and T_2 is in effect the time taken for a spin to precess once about the local field. If any interaction with the spins occurs in a time less than T_2, for example a reversal of the field, then the spins will not have any opportunity to redistribute their energy or to follow the change in direction in such a short time interval.

B

The magnetization of the sample M is simply the nett effect of all the individual moments μ_i so that

$$M = \sum_{\substack{\text{Unit} \\ \text{vol}}} \mu_i$$

and, classically, we may write

$$\frac{dM}{dt} = \gamma(\mathbf{H} \times \mathbf{M}) \qquad (1.1)$$

It is usual to take H_0 to be in the z direction, and H_1 in the xy plane. Bloch introduced the relaxation times T_1 and T_2 into the theory by writing these into the components of equation (1.1) in the following way

$$\frac{dM_z}{dt} = \gamma(\mathbf{H} \times \mathbf{M})_z + \frac{(M_0 - M_z)}{T_1} \qquad (1.2a)$$

$$\frac{dM_x}{dt} = \gamma(\mathbf{H} \times \mathbf{M})_x - \frac{M_x}{T_2} \qquad (1.2b)$$

$$\frac{dM_y}{dt} = \gamma(\mathbf{H} \times \mathbf{M})_y - \frac{M_y}{T_2} \qquad (1.2c)$$

These are the so-called Bloch equations. Equation $(1.2a)$ shows that, after the removal of the radio-frequency field H_1, the z component of $\mathbf{M}$ grows as the individual moments μ_i flip to bring M_z towards its equilibrium value M_0. This process requires the moments to give up energy to the lattice and is characterized by the spin-lattice relaxation time T_1. Since T_1 appears only in the z component, it is sometimes called the 'longitudinal' relaxation time.

Similarly we can see that T_2 refers to the rate at which the x and y components of $\mathbf{M}$ decay to their equilibrium zero value, and T_2 therefore characterizes the perpendicular component of magnetization, and is sometimes referred to as the 'transverse' relaxation time. The steady state solution of these Bloch equations (i.e. when $dM_z/dt = 0$) in the presence of an oscillating field $2H_1 \exp(i\omega t)$ is well known (see, for example, Pake[7]) and, if we define in the usual way a complex susceptibility, $\chi = \chi' - i\chi''$, we obtain

$$\chi' = \tfrac{1}{2}\chi_0 \omega_0 T_2 \frac{T_2(\omega_0 - \omega)}{1 + T_2{}^2(\omega_0 - \omega)^2 + \gamma H_1{}^2 T_1 T_2} \qquad (1.3a)$$

and

$$\chi'' = \tfrac{1}{2}\chi_0\omega_0 T_2 \frac{1}{1 + T_2{}^2(\omega_0 - \omega)^2 + \gamma^2 H_1{}^2 T_1 T_2} \qquad (1.3b)$$

where the static susceptibility $\chi_0 = M_0/H_0$.

The mean rate at which energy is absorbed from the radiation field is given by

$$Z = \overline{H_x \frac{dM_x}{dt}} = \frac{\omega}{2\pi} \int\limits_0^{2\pi/\omega} \mathbf{H}_0 \cdot \frac{d\mathbf{M}}{dt}\, dt = 2\omega\chi'' H_1{}^2$$

$$= \frac{\omega\omega_0\chi_0\, T_2 H_1{}^2}{1 + T_2{}^2(\omega_0 - \omega)^2 + \gamma^2 H_1{}^2 T_1 T_2} \qquad (1.4)$$

and if H_1 is small, then $\gamma^2 H_1{}^2 T_1 T_2 \ll 1$ and

$$Z = \frac{\omega\omega_0\chi_0 T_2 H_1{}^2}{1 + T_2{}^2(\omega_0 - \omega)^2} \qquad (1.5)$$

A graph of absorption Z against ω (or against $\omega_0 = \gamma H_0$) defines the usual resonance curve. The shape of this curve is Lorentzian, and this arises, as in all damped harmonic oscillator problems, because of the assumption of an exponential decay caused by relaxation. However, lines of a more general shape than this are commonly observed. The Lorentzian curve reaches a maximum value at $\omega = \omega_0$, and the half-width at half maximum value is given by

$$\Delta\omega_{\frac{1}{2}} = \frac{1}{T_2} = \gamma\Delta H_{\frac{1}{2}}$$

where $\Delta H_{\frac{1}{2}}$ is the half-width in units of magnetic field. (In an E.S.R. experiment it is more usual to vary the field through the resonance condition and plot power absorption as a function of field.)

It can also be seen from equation (1.4) that for frequencies near resonance, i.e. as $\omega \to \omega_0$, the height of the resonance curve falls as H_1 is increased. This phenomenon is called 'saturation' of the spin system, and leads to a method of measuring T_1, which is discussed in more detail in Chapter 6.

The rate equation approach

Consider a crystal of volume V containing a total number of N spins (paramagnetic ions), immersed in a liquid helium bath at a temperature T and placed in a magnetic field H_0. We shall assume that the energy-level splitting Δ_c is sufficiently great that only the lowest two levels are appreciably populated. This is often, but by no means always, the case in practice; complications due to other populated levels will be discussed later. This two-level system is depicted in Fig. 1.2. Levels a and b, between which we are observing paramagnetic resonance, have thermal equilibrium populations N_a and N_b respectively, where

$$N_a + N_b = N \tag{1.6}$$

and we define

$$n_0 = N_a - N_b \tag{1.7}$$

as the population difference in thermal equilibrium.

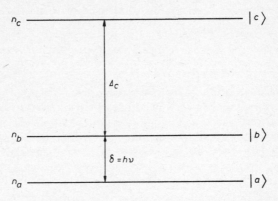

FIG. 1.2. Simple energy-level diagram for an ion in a crystalline field (splitting Δ_c) and also in a magnetic field (splitting δ) where $\Delta_c \gg \delta$.

N_a and N_b are determined by the Boltzmann equation

$$\frac{N_b}{N_a} = \exp(-\delta/kT) \tag{1.8}$$

where $\delta = E_b - E_a = h\nu$ (see Fig. 1.2).

If the system is disturbed in any way, the non-equilibrium populations are given by n_a and n_b where

$$n_a + n_b = N \tag{1.9}$$

$$n = n_a - n_b \tag{1.10}$$

and

$$\frac{n_b}{n_a} = \exp(-\delta/kT_s) \tag{1.11}$$

This defines a spin temperature T_s, at least for our simple two-level case, as the effective temperature of the disturbed spin system. In thermal equilibrium this becomes just the temperature of the crystal lattice T.

The disturbed spin system will tend to regain its thermal equilibrium distribution due to interactions with the lattice, whose thermal vibrations we refer to as the 'phonon system'. This consists of a set of lattice oscillators, each with average energy $\bar{E} = (p+\frac{1}{2})\delta$, where p is the average phonon excitation number, used to characterize the state of the phonon system.

The density of phonon modes at a frequency v is assumed to have its classical value

$$\rho(\delta)d(\delta) = \frac{3V\delta^2}{2\pi^2\hbar^3\,v^3}\,d(\delta) \qquad (1.12)$$

where v is the velocity of sound in the crystal and $\delta = hv$. If we assume that the phonons are coupled tightly to the bath then the phonon excitation will always have its thermal equilibrium value

$$P(\delta) = [\exp(\delta/kT)-1]^{-1} \qquad (1.13)$$

Defining $W_{a\to b}$ as the probability per second for a spin to go from level a to level b, and $W_{b\to a}$ as the probability per second for the reverse process, we can write the change in population of level a as

$$\frac{dn_a}{dt} = W_{b\to a}n_b - W_{a\to b}n_a \qquad (1.14)$$

with a similar expression for level b.

The rate equation for the population difference n is given by

$$\frac{dn}{dt} = \frac{dn_a}{dt} - \frac{dn_b}{dt} = 2(W_{b\to a}n_b - W_{a\to b}n_a)$$

$$= (W_{b\to a} - W_{a\to b})N - (W_{b\to a} + W_{a\to b})n \qquad (1.15)$$

and if we assume that the spins are coupled to the phonons by a time-dependent relaxation perturbation (the nature of which we shall discuss in Chapter 2) then we can use as the matrix elements for phonon creation and absorption those for the harmonic oscillator. This enables us to write

$$W_{b\to a} = K[P(\delta)+1]\ \mathrm{sec}^{-1}$$

and

$$W_{a\to b} = KP(\delta)\ \mathrm{sec}^{-1} \qquad (1.16)$$

where K is some factor independent of temperature. Hence

$$\frac{dn}{dt} = 2K[-n_bP(\delta) - n_b + n_aP(\delta)] \qquad (1.17)$$

where the three terms on the right represent stimulated emission, spontaneous emission, and absorption of phonons, respectively.

From equations (1.6), (1.7) and (1.8) we can show that

$$n_0 = N \tanh(\delta/2kT) \qquad (1.18)$$

hence from equation (1.17)

$$\frac{dn}{dt} = K\{N - [2P(\delta) + 1]n\}$$

and using equations (1.13) and (1.18)

$$\frac{dn}{dt} = K \coth(\delta/2kT)(n_0 - n) \qquad (1.19)$$

Thus defining

$$T_1 = \frac{1}{W_{a \to b} + W_{b \to a}} = \frac{1}{K \coth(\delta/2kT)} \qquad (1.20)$$

we obtain

$$\frac{dn}{dt} = \frac{n_0 - n}{T_1} \qquad (1.21)$$

If $\delta \ll 2kT$, then (1.20) simplifies to

$$T_1^{-1} = 2kT/\delta \equiv AT$$

which is the temperature dependence for the direct process, since we have considered an energy-conserving process involving one spin transition and one phonon.

Equation (1.21) has the solution

$$n = n_0[1 - \exp(-t/T_1)] \qquad (1.22)$$

which shows that, after being disturbed, the rate at which the spin system regains its thermal equilibrium distribution follows an exponential law, characterized by the time constant T_1. We shall make use of this result in Chapter 7 when discussing methods of measuring T_1.

At very low temperatures, where $2kT \ll h\nu$, the relaxation rate approaches the temperature independent value $T_1^{-1} = K$, which is just the transition probability for spontaneous emission of phonons. It has so far been sufficient simply to define K as a temperature independent factor. In Chapter 2 we shall see how to calculate the value of this constant from crystal-field parameters.

A similar argument can be used to deduce an expression for T_1 when relaxation occurs by means of a two-phonon process. Consider the Orbach process outlined on p. 5 in which relaxation between levels b and a takes place via level c (Fig. 1.2). We may write the transition probabilities for the $b-c$ transition, following equation (1.16), as

$$W_{b \to c} = K_1 P(\Delta_c) \text{ sec}^{-1}$$

$$W_{c \to b} = K_1 [P(\Delta_c) + 1] \text{ sec}^{-1}$$

and similarly for the $a-c$ transition as

$$W_{a \to c} = K_2 P(\Delta_c + \delta) \text{ sec}^{-1}$$

$$W_{c \to a} = K_2 [P(\Delta_c + \delta) + 1] \text{ sec}^{-1}$$

Assuming that there is no direct relaxation between levels a and b, the rate equations become, as in equation (1.14),

$$\frac{dn_a}{dt} = K_0 \{ n_c [P(\Delta_c + \delta) + 1] - n_b P(\Delta_c + \delta) \} \quad (1.23a)$$

and

$$\frac{dn_b}{dt} = K_0 \{ n_c [P(\Delta_c) + 1] - n_b P(\Delta_c) \} \quad (1.23b)$$

where

$$K_0 = \frac{2K_1 K_2}{(K_1 + K_2)}$$

The rate equation for the population difference n is now given by

$$\frac{dn}{dt} = \frac{dn_a}{dt} - \frac{dn_b}{dt}$$

$$= K_0 [n_c P(\Delta_c) - n_b P(\Delta_c) - n_c P(\Delta_c + \delta) + n_a P(\Delta_c + \delta)] \quad (1.24)$$

If $kT \ll \Delta_c$, n_c is negligible compared with n_a and n_b, and equation (1.24) becomes

$$\frac{dn}{dt} = K_0 [n_a P(\Delta_c + \delta) - n_b P(\Delta_c)]$$

$$= \tfrac{1}{2} K_0 [(n + N) P(\Delta_c + \delta) - (N - n) P(\Delta_c)] \quad (1.25)$$

The phonon excitation numbers $P(\Delta_c)$ and $P(\Delta_c + \delta)$ are given by expressions similar to equation (1.13), and if $\Delta_c \gg kT$ we may write

$$P(\Delta_c) \approx \exp(-\Delta_c / kT)$$

and

$$P(\Delta_c + \delta) \approx \exp[-(\Delta_c + \delta)/kT]$$

Combining equations (1.6), (1.7) and (1.8), we may show that

$$\exp(-\delta/kT) = \frac{N-n_0}{N+n_0}$$

and hence finally

$$\frac{dn}{dt} = \frac{n_0-n}{T_1(O)} \tag{1.26}$$

where

$$T_1(O)^{-1} = K_0 \frac{N}{N+n_0} P(\Delta_c) \approx K_0 \exp(-\Delta_c/kT)$$

K_0 will also be evaluated in Chapter 2.

Expressions for T_1 are calculated explicitly in Chapter 2 for the simple two-level case, and the complicating effects of cross relaxation and the phonon bottleneck are each discussed at length in Chapters 3 and 4 respectively. Chapters 5 to 8 are devoted to a critical appraisal of the various experimental techniques for measuring spin-lattice relaxation times, classified under the broad headings of non-resonant and resonant methods. The latter is further subdivided into three: continuous saturation, pulse saturation and monitored recovery, and other resonant methods, the last two categories differing not so much in principle as in practical detail. In Chapter 9 and Appendix 2 an attempt is made to accumulate some of the more pertinent experimental results and to interpret these in the spirit of the earlier sections.

References

1.   Casimir, H. B. G. and du Pré, F. K., *Physica*, 1938, **4,** 579.
2.   Van Vleck, J. H., *J. Chem. Phys.*, 1939, **7,** 72.
3.   Van Vleck, J. H., *Phys. Rev.*, 1940, **57,** 426.
4.   Assenheim, H. M., *Introduction to Electron Spin Resonance* (Adam Hilger Ltd, 1966; Plenum Press, New York, 1967).
5.   Orbach, R., *Proc. Roy. Soc.*, 1961, A**264,** 458.
6.   Bloch, F., *Phys. Rev.*, 1946, **70,** 460.
7.   Pake, G. E., *Paramagnetic Resonance* (W. A. Benjamin Inc., 1962).

Chapter 2

Theoretical Description of Spin-Lattice Relaxation

The Two-Level Case

The first paper on this subject was by Waller[1] in 1932. His calculations, appearing before any of the relaxation measurements of Gorter[2] and his co-workers, involved both spin-spin and spin-lattice interactions in paramagnetic salts. Waller considered that the mechanism by which the spin system relaxed to the lattice was essentially through modulation of the inter-ionic magnetic dipolar interaction, as a result of lattice vibrations. That is to say, the lattice vibrations only influenced a spin through the modulation of the spin-spin interaction. The predicted variations of T_1 with temperature were of the form later observed experimentally, but Waller's calculated values of T_1 were several orders of magnitude greater than those measured later by Gorter.

Heitler and Teller[3] in 1936, and Fierz[4] in 1938, attempted to modify the Waller calculations, in what Van Vleck later described as a schematic manner, by the addition of an interaction arising from the modulation of the *crystalline Stark effect* by the lattice vibrations. Although this reduced the discrepancy between theory and experiment, the calculated orders of magnitude were still too high by factors between 100 and 10 000.

In 1939 Kronig[5] published what was essentially a phenomenological treatment of spin-lattice relaxation based upon this same premise, namely that 'when the lattice vibrates, the orbital motion undergoes periodic changes due to variations in the electric field of the crystal and these changes . . . react upon the spins causing them to alter their orientation with respect to the constant external magnetic field'. Kronig's treatment appeared at roughly the same time as a paper due to Van Vleck[6] giving a detailed calculation of the Jahn-Teller effect and crystalline Stark splitting for clusters of ions of the form XY_6. A year later using the results of the first paper, Van Vleck[7] published his pioneer paper on paramagnetic relaxation times in titanium and chrome alum; it is significant that the forms of the dependence of the relaxation times upon temperature were similar to those predicted by Kronig. By this time the experimental results of Gorter and his co-workers were available and in Van Vleck's own words, the agreement between experiment and theory was 'miserable'. This was an unfortunate expression, correct at the time when referring to the alums, but the fault is now known to lie in the behaviour of the materials themselves rather than in the fundamental tenets of the theory. In recent years further calculations have been carried out, based on this same assumption that the dominant relaxation mechanism is one in which the lattice vibrations within the real crystal modulate the crystalline electric field acting upon the magnetic ion in question and thereby induce spin transitions. There is no direct interaction between the crystal field of the lattice and the spin, but there is between the lattice and the orbital motion of the electron, and the spin is made aware of this through the spin-orbit interaction. This model is particularly valid when the magnetic ions are diluted (i.e. well separated) in the lattice; modulation of the exchange interaction can be important when dealing with clusters of ions, and this is essentially the Waller mechanism.

In an article on spin-lattice relaxation, Stevens[8] noted that as the crystal-field theory has been increasingly applied, so it has become clear that its success has been mainly due to the fact that the crystal-field parameters are chosen to fit observations; a model relying solely upon electrostatic fields is too restrictive and more attention is at present being given by theoreticians to molecular-orbital-type models. These models, at the time of writing, have not made significant progress and the crystal-field approach is given below in default of any other more elegant and convincing solution. Quite possibly a more accurate theory based upon the molecular orbital approach will eventually be forthcoming. However, symmetry considerations usually decide the number of free parameters required, and will have to be included in any new theory. Thus the form of many of the crystal-field theory results will be retained, although with different matrix elements.

Outline of the crystal-field modulation model

The modulation of the crystal field occurs because the lattice vibrations vary the interatomic spacings of the atoms or ions in the lattice in a manner which is time dependent. The crystal field reflects this time dependence. As Stevens[8] observed, a truly instantaneous picture of the lattice would reveal many distortions from the regular ideal lattice symmetry. Clearly the rate at which that picture changes its form will depend upon the frequencies of the lattice vibration present.

Obviously, a paramagnetic ion A in a crystal will have a number of neighbours, each of which produces its own potential at A. It is convenient to consider the potential due to each neighbour separately, and then to proceed to the appropriate summation.

Consider a set of axes with origin O at A and Oz directed towards the particular neighbour B we are considering. If A is displaced it carries O with it, so the co-ordinates (x, y, z) of an electron in A are unchanged. But the potential seen by this electron has changed, since it has altered its position with respect to B. Similarly the potential at (x, y, z) is changed if B moves. Let A be displaced by (X_A, Y_A, Z_A) and B by (X_B, Y_B, Z_B). Then the potential due to B which the electron experiences is that which it would have found at the position $(x - X_B + X_A,\ y - Y_B + Y_A,\ z - Z_B + Z_A)$ had there been no displacements.

Writing $V_\alpha(x, y, z)$ and $V_\beta(x, y, z)$ as the potential energies before and after displacement, we have

$$V_\beta(x, y, z)$$

$$= V(x - X_B + X_A,\ y - Y_B + Y_A,\ z - Z_B + Z_A)$$

$$= V_\alpha(x, y, z) + (X_A - X_B)\frac{\partial V}{\partial x} + (Y_A - Y_B)\frac{\partial V}{\partial y} + (Z_A - Z_B)\frac{\partial V}{\partial z}$$

$$+ \frac{1}{2!}\left[\begin{array}{l} (X_A - X_B)^2 \dfrac{\partial^2 V}{\partial x^2} + (Y_A - Y_B)^2 \dfrac{\partial^2 V}{\partial y^2} + \cdots \\[2mm] + 2(X_A - X_B)(Y_A - Y_B)\dfrac{\partial^2 V}{\partial x \partial y} + \cdots \end{array} \right] \quad (2.1)$$

Thus the change in potential energy can be calculated by forming derivatives of the 'static' potential $V(x, y, z)$ with respect to (x, y, z).

The relative displacements of the neighbours surrounding a magnetic ion give rise to calculable changes in the potential experienced by that ion. Such changes will occur on account of thermal and zero-point vibrations and it is the purpose of spin-lattice relaxation calculations to determine the transition probabilities which arise from such temporal modulation of the crystal fields. We outline below two forms of the calculation; it is to be stressed that the basic physical model in each is that of Van Vleck and Kronig and that the two approaches differ primarily in their mathematical formulation of the problem.

§2.1

THE ORBACH-SCOTT AND JEFFRIES CALCULATIONS
USING OPERATOR EQUIVALENTS

These calculations apply particularly to the rare-earth ions and they have had most encouraging success, especially in explaining the experimental data of Scott and Jeffries.[9] Below, the crystal-field model is developed further and the reader is reminded of the operator-equivalent technique, before the relaxation calculations are outlined.

In equation (2.1) we obtained a formula for the change in potential energy of an ion due to spatial displacements. Introducing the polar co-ordinates (r, θ, ϕ), we write the potential energy as

$$V(x, y, z) = \sum_{n \neq 0} A_n{}^m r^n Y_n{}^m (\theta, \phi) \qquad (2.2)$$

where each $A_n{}^m$ is a constant determined by the charge distribution and the separation of the neighbour B, r^n the (mean) nth power of the distance from the nucleus of the magnetic electron (usually a d-electron for transition metal ions and an f-electron for rare-earth ions) and $Y_n{}^m$ the spherical harmonic of degree n and azimuthal quantum number m (the reader is referred to a paper by Hutchings[10] which sets out the situation of the crystal-field theory as it stood in 1964).

The parameters $Y_n{}^m(\theta, \phi)$ are defined in Hutchings paper as

$$Y_n{}^m(\theta, \phi) = (-1)^{(m + |m|)/2} \left[\frac{(2n + 1)\,(n - |\,m\,|)!}{2(n + |\,m\,|)!} \right]^{\frac{1}{2}} \times$$

$$\frac{1}{(2\pi)^{\frac{1}{2}}} P_n{}^{|m|} (\cos\,\theta)\,\exp(im\phi)$$

where

$$P_n{}^0(\mu) = \frac{1}{2^n n!} \cdot \frac{d^n(\mu^2 - 1)^n}{d\mu^n} \quad (\mu = \cos\,\theta)$$

and

$$P_n{}^{|m|}(\mu) = (1 - \mu^2)^{|m|/2} \cdot \frac{d^{|m|}[P_n{}^0(\mu)]}{d\mu^{|m|}}$$

It is usual to regard the $A_n{}^m$ as parameters chosen to fit experimental results, although they may be calculated for particular models. Their dimensions are energy/(distance)^n.

Stevens gives the following formulae which are useful in forming the derivatives of V.

$$\frac{\partial}{\partial x}\left(r^n Y_n^{\ 0}\right) = \left[\frac{(2n+1)\ n(n-1)}{2n-1}\right]^{\frac{1}{2}} r^{n-1}\left(\frac{Y_{n-1}^{(1)} - Y_{n-1}^{(-1)}}{2}\right)$$

$$\frac{\partial}{\partial y}\left(r^n Y_n^{\ 0}\right) = \left[\frac{(2n+1)\ n(n-1)}{2n-1}\right]^{\frac{1}{2}} r^{n-1}\left(\frac{Y_{n-1}^{(1)} + Y_{n-1}^{(-1)}}{2i}\right)$$

$$\frac{\partial}{\partial z}\left(r^n Y_n^{\ 0}\right) = \left(\frac{2n+1}{2n-1}\right)^{\frac{1}{2}} n r^{n-1} Y_{n-1}^0$$

and for $m \geqslant 1$

$$\frac{\partial}{\partial x}\left(r^n Y_n^{\ m}\right) = -\left[\frac{(2n+1)\ (n+m)\ (n+m-1)}{2n-1}\right]^{\frac{1}{2}} \tfrac{1}{2}r^{n-1} Y_{n-1}^{m-1}$$

$$+ \left[\frac{(2n+1)\ (n-m)\ (n-m-1)}{2n-1}\right]^{\frac{1}{2}} \tfrac{1}{2}r^{n-1} Y_{n-1}^{m+1}$$

$$\frac{\partial}{\partial y}\left(r^n Y_n^{\ m}\right) = -i\left[\frac{(2n+1)\ (n+m)\ (n+m-1)}{2n-1}\right]^{\frac{1}{2}} \tfrac{1}{2}r^{n-1} Y_{n-1}^{m-1}$$

$$-i\left[\frac{(2n+1)\ (n-m)\ (n-m-1)}{2n-1}\right]^{\frac{1}{2}} \tfrac{1}{2}r^{n-1} Y_{n-1}^{m+1}$$

$$\frac{\partial}{\partial z}\left(r^n Y_n^{\ m}\right) = \left[\frac{(2n+1)\ (n+m)\ (n-m)}{2n-1}\right]^{\frac{1}{2}} r^{n-1} Y_n^m {}_1$$

The results for $m \leqslant -1$ are obtained by taking complex conjugates. The spherical harmonics above are in the forms for conversion into equivalent operators (see below) and many of these have been tabulated by Hutchings.

In most relaxation calculations it suffices to neglect matrix elements to excited configurations of opposite parity and then all odd-parity crystal-field-like terms have zero matrix elements and are omitted. Moreover, with the iron group ions all terms with n greater than 5 vanish and for the rare earths they do so beyond $n = 7$. Also, the term with $n = 1$ does not involve (x, y, z) and therefore does not contribute to the relaxation.

If we confine our attention to rare-earth salts, the electronic states may be given in terms of the appropriate $\mathscr{J}$ value. From the work of Stevens[11] it is possible to represent any crystal-field operator acting on the orbital moment as an equivalent operator acting on states of total angular momentum $\mathscr{J}$. Hence the modulations of the crystal field, acting only on the orbital part of $\mathscr{J}$ in the limit of strong spin-orbit coupling, can be transformed to operate directly on states of $\mathcal{J}$ by the use of multiplying factors tabled by Elliott and Stevens.[12] As an example of this let us take the case quoted by Orbach.[13] He considers the trivalent rare-earth ion in the ethyl sulphates and writes the crystalline field potential V in terms of the spherical harmonics as

$$
\begin{aligned}
V = \sum_{n,m} V_n{}^m = \sum_{n,m} & A_n{}^m r^n Y_n{}^m(\theta, \phi) \\
= & A_2{}^0(3z^2 - r^2) + A_4{}^0(35z^4 - 30r^2z^2 + 3r^4) \\
& + A_6{}^0(231z^6 - 215z^4r^2 + 105z^2r^4 - 5r^6) \\
& + A_6{}^6(x^6 - 15x^4y^2 + 15x^2y^4 - y^6)
\end{aligned}
$$

where the last term is a linear combination of $Y_6{}^{+6}$ and $Y_6{}^{-6}$. (The $A_n{}^m$'s required are determined from symmetry considerations. Hutchings notes that for the C_{3h} symmetry of ethyl sulphates $A_2{}^0$, $A_4{}^0$, $A_6{}^0$, $A_6{}^6$ and $A_6{}^{-6}$ may be required for values of n up to 6; for the C_{3v} symmetry of double nitrates, the terms are $A_2{}^0$, $A_4{}^0$, $A_4{}^3$, $A_6{}^0$, $A_6{}^3$ and $A_6{}^6$.) Using the concept of the equivalent operator, referred to above, this potential may be rewritten as

$$
\begin{aligned}
V = A_2{}^0 r^2 \langle \mathcal{J} \| \alpha \| \mathcal{J} \rangle O_2{}^0 + A_4{}^0 r^4 \langle \mathcal{J} \| \beta \| \mathcal{J} \rangle O_4{}^0 \\
+ A_6{}^0 r^6 \langle \mathcal{J} \| \gamma \| \mathcal{J} \rangle O_6{}^0 + A_6{}^6 r^6 \langle \mathcal{J} \| \gamma \| \mathcal{J} \rangle O_6{}^6
\end{aligned}
$$

where there is term-by-term equivalence with the above equation and the $O_n{}^m$ are given in the Table 2.1. The coefficients, $\langle \mathcal{J} \| \alpha \| \mathcal{J} \rangle = \alpha_J$, $\langle \mathcal{J} \| \beta \| \mathcal{J} \rangle = \beta_J$ and $\langle \mathcal{J} \| \gamma \| \mathcal{J} \rangle = \gamma_J$, are

TABLE 2.1. Form of operators $O_n{}^m$

$$
\begin{aligned}
O_2^0 =\ & 3\mathcal{J}_z{}^2 - \mathcal{J}(\mathcal{J}+1) \\
O_2^1 =\ & \tfrac{1}{4}[\mathcal{J}_z(\mathcal{J}_+ + \mathcal{J}_-) + (\mathcal{J}_+ + \mathcal{J}_-)\mathcal{J}_z] \\
O_2^2 =\ & \tfrac{1}{2}[\mathcal{J}_+{}^2 + \mathcal{J}_-{}^2] \\
O_4^0 =\ & 35\mathcal{J}_z{}^4 - [30\mathcal{J}(\mathcal{J}+1) - 25]\mathcal{J}_z{}^2 - 6\mathcal{J}(\mathcal{J}+1) + 3\mathcal{J}^2(\mathcal{J}+1)^2 \\
O_4^1 =\ & \tfrac{1}{4}\{[7\mathcal{J}_z{}^3 - 3\mathcal{J}(\mathcal{J}+1)\mathcal{J}_z - \mathcal{J}_z](\mathcal{J}_+ + \mathcal{J}_-) + (\mathcal{J}_+ + \mathcal{J}_-)[7\mathcal{J}_z{}^3 - 3\mathcal{J}(\mathcal{J}+1)\mathcal{J}_z - \mathcal{J}_z]\} \\
O_4^2 =\ & \tfrac{1}{4}\{[7\mathcal{J}_z{}^2 - \mathcal{J}(\mathcal{J}+1) - 5](\mathcal{J}_+{}^2 + \mathcal{J}_-{}^2) + (\mathcal{J}_+{}^2 + \mathcal{J}_-{}^2)[7\mathcal{J}_z{}^2 - \mathcal{J}(\mathcal{J}+1) - 5]\} \\
O_4^3 =\ & \tfrac{1}{4}[\mathcal{J}_z(\mathcal{J}_+{}^3 + \mathcal{J}_-{}^3) + (\mathcal{J}_+{}^3 + \mathcal{J}_-{}^3)\mathcal{J}_z] \\
O_4^4 =\ & \tfrac{1}{2}[\mathcal{J}_+{}^4 + \mathcal{J}_-{}^4] \\
O_6^0 =\ & 231\mathcal{J}_z{}^6 - 105[3\mathcal{J}(\mathcal{J}+1) - 7]\mathcal{J}_z{}^4 + [105\mathcal{J}^2(\mathcal{J}+1)^2 - 525\mathcal{J}(\mathcal{J}+1) + 294]\mathcal{J}_z{}^2 \\
& - 5\mathcal{J}^3(\mathcal{J}+1)^3 + 40\mathcal{J}^2(\mathcal{J}+1)^2 - 60\mathcal{J}(\mathcal{J}+1) \\
O_6^1 =\ & \tfrac{1}{4}\{[33\mathcal{J}_z{}^5 - \{30\mathcal{J}(\mathcal{J}+1) - 15\}\mathcal{J}_z{}^3 + \{5\mathcal{J}^2(\mathcal{J}+1)^2 - 10\mathcal{J}(\mathcal{J}+1) + 12\}]\mathcal{J}_z(\mathcal{J}_+ + \mathcal{J}_-) \\
& + (\mathcal{J}_+ + \mathcal{J}_-)[33\mathcal{J}_z{}^5 - \{30\mathcal{J}(\mathcal{J}+1) - \text{etc.}\}]\} \\
O_6^2 =\ & \tfrac{1}{4}\{[33\mathcal{J}_z{}^4 - \{18\mathcal{J}(\mathcal{J}+1) + 123\{\mathcal{J}_z{}^2 + \mathcal{J}^2(\mathcal{J}+1)^2 + 10\mathcal{J}(\mathcal{J}+1) + 102](\mathcal{J}_+{}^2 + \mathcal{J}_-{}^2) \\
& + (\mathcal{J}_+{}^2 + \mathcal{J}_-{}^2)[33\mathcal{J}_z{}^4 - \{18\mathcal{J}(\mathcal{J}+1) + \text{etc.}\}]\} \\
O_6^3 =\ & \tfrac{1}{4}\{[11\mathcal{J}_z{}^3 - 3\mathcal{J}(\mathcal{J}+1)\mathcal{J}_z - 59\mathcal{J}_z](\mathcal{J}_+{}^3 + \mathcal{J}_-{}^3) + (\mathcal{J}_+{}^3 + \mathcal{J}_-{}^3)[11\mathcal{J}_z{}^3 - 3\mathcal{J}(\mathcal{J}+1)\mathcal{J}_z \\
& - 59\mathcal{J}_z]\} \\
O_6^4 =\ & \tfrac{1}{4}\{[11\mathcal{J}_z{}^2 - \mathcal{J}(\mathcal{J}+1) - 38](\mathcal{J}_+{}^4 + \mathcal{J}_-{}^4) + (\mathcal{J}_+{}^4 + \mathcal{J}_-{}^4)[11\mathcal{J}_z{}^2 - \mathcal{J}(\mathcal{J}+1) - 38]\} \\
O_6^5 =\ & \tfrac{1}{4}[\mathcal{J}_z(\mathcal{J}_+{}^5 + \mathcal{J}_-{}^5) + (\mathcal{J}_+{}^5 + \mathcal{J}_-{}^5)\mathcal{J}_z] \\
O_6^6 =\ & \tfrac{1}{2}[\mathcal{J}_+{}^6 + \mathcal{J}_-{}^6]
\end{aligned}
$$

given by Elliott and Stevens.[14] Using the above value for V and coefficients $A_n{}^m r^n$ given by Baker and Bleaney[15] for holmium ethyl sulphate, Orbach derives the following wave functions and splitting within the ground multiplet for this particular salt.

$g_\parallel$	$g_\perp$	Zero-field splitting (cm^{-1})	Wave function
7.705 ± 0.005	~ 4	—	$0.933\,\lvert \pm 7 \rangle + 0.342\,\lvert \pm 1 \rangle$
0	0	5.55 ± 0.5	$0.5[\,\lvert 6 \rangle + \lvert -6 \rangle\,] + 0.707\,\lvert 0 \rangle$

To sum up so far, we have shown how the change in potential due to the movement of a near neighbour can be calculated in terms of derivatives of the static potential. By a brief reference to the work of Stevens and an example quoted from Orbach we have shown how the potential problem becomes more tractable through the use of operator equivalents. Returning to the relaxation problem, naturally we have to take account of all the neighbours and not just the single one referred to above as B. We must convert the various combinations of the displacements in the general case into normal modes of the ion A plus all its neighbours (i.e. normal modes of a cluster of centres, not normal modes of the lattice), and hence, taking into account the strains which these displacements cause, we must find the contribution $\mathscr{H}_0$ from this cause to the dynamic Hamiltonian of the system. Stevens[8] tackles this problem in a general form. We shall adopt and adapt the more limited approach of Orbach and of Scott and Jeffries[9] relating specifically to rare-earth ions.

We have already shown how an ionic displacement causes a change in the potential energy, and, as in equation (2.2), we may write generally a static crystal-field interaction, as introduced by Stevens.

Now the perturbation which causes the relaxation is just the fluctuation in the static crystal-field interaction due to thermal lattice strains ε. Strictly, the components of a strain tensor should be included in order that anisotropy of the elastic waves in the lattice may be taken into account, but, in practice, it usually suffices to include ε as an average strain and to ignore directional properties.

We can find the form of the required interaction term as follows. Let ξ represent the co-ordinate of a neighbouring ion; we expand $A_n{}^m(\xi)$ in a Taylor series as

$$A_n{}^m(\xi) = A_n{}^m(\xi)_0 + \varepsilon\xi \left(\frac{\partial A_n{}^m}{\partial \xi} \right)_0$$

$$+ \text{second order terms in } \varepsilon^2 \xi^2$$

$$+ \ldots \ldots \tag{2.3}$$

The first term on the right-hand side of this equation represents the static condition and does not give rise to relaxation. The term in ε gives rise to the single phonon direct relaxation process, while terms in ε^2 contribute to the Raman process (see below).

Scott and Jeffries note that if $A_n{}^m$ is assumed to be proportional to some inverse power of ξ, it is permissible to take as an approximation

$$\left| \xi \left(\frac{\partial A_n{}^m}{\partial \xi} \right)_0 \right| \approx \left| A_n{}^m(\xi)_0 \right|$$

where the right-hand term is just the static one, which is often known experimentally.

Thus we may write down for the first-order relaxation process —the direct process discussed in the next paragraph—an expression for the orbit-lattice interaction V_{ol} which causes the relaxation, namely

$$V_{ol}(j) = \varepsilon \sum_{n,m} A_n{}^m \mathbf{r}(j)^n Y_n{}^m = \varepsilon \sum_{n,m} V_n{}^m$$

$$= \varepsilon \sum_{n,m} A_n{}^m \mathbf{r}(j)^n \langle \mathcal{J} \| x_n \| \mathcal{J} \rangle O_n{}^m(j) \qquad (2.4)$$

where the expression is written for the jth ion, $\mathbf{r}(j)$ is the radius vector to the jth paramagnetic ion site from the arbitrary origin and $\langle \mathcal{J} \| x_n \| \mathcal{J} \rangle = \alpha$, β or γ depending on whether $n = 2, 4$ or 6 respectively.

The direct or one-phonon process

As explained in Chapter 1, this situation predominates at low temperatures and is one in which the relaxation of the spin system takes place as a result of a spin flip from state $| b \rangle$ to state $| a \rangle$ with the simultaneous creation of a phonon of energy $E_b - E_a = \delta_{ab}$, thus ensuring the conservation of energy. A hypothetical energy-level diagram for this system is shown in Fig. 1.2. In equation (1.11), we assumed populations n_b and n_a in the states $| b \rangle$ and $| a \rangle$ and we defined a spin temperature T_s through

$$\frac{n_b}{n_a} = \exp(-\delta_{ab}/kT_s)$$

In thermal equilibrium $T_s = T$ the lattice temperature. We assume the lattice to be in close thermal contact with a bath of infinite thermal capacity at temperature T_b. T_b is therefore constant and normally equals T. However, conditions can arise when this is not so and we shall take account of this possibility in Chapter 4. For the present we shall assume $T_b = T$ the temperature of the lattice.

If the state of the system is disturbed so that T_s becomes greater than T, the orbit-lattice interaction will induce transitions between the states $|b\rangle$ and $|a\rangle$ and we require to calculate the rate at which the spin system returns to equilibrium with the lattice by means of the emission or absorption of a single phonon. The problem is shown diagrammatically in Fig. 2.1. When a single transition is made from level b to level a, energy is conserved by the number of phonons P_{nm} in the mode n,m increasing by unity.

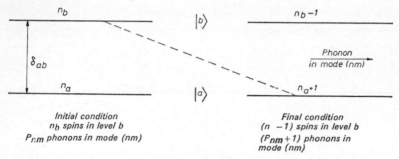

FIG. 2.1. Diagram of the direct process. When the single spin transition from $|b\rangle$ to $|a\rangle$ is made, a single phonon of energy δ_{ab} is emitted.

From first-order time-dependent perturbation theory (see, for example, Schiff[16]), the probability per second of making a transition between b and a is given in these conditions by

$$W_{b \to a} = \frac{2\pi}{\hbar} \left| \langle P_{nm} | \varepsilon | P_{nm} + 1 \rangle \right|^2 \left| \langle a | \sum_{n,m} V_n{}^m | b \rangle \right|^2 \rho(\delta) \tag{2.5}$$

where the transition is from state $|b, P_{nm}\rangle$ to $|a, P_{nm} + 1\rangle$.

As explained in Chapter 1, p. 11 the occupation number of the phonon mode is assumed to have its thermal equilibrium value

$$P_{nm} = [\exp(\delta_{ab}/kT) - 1]^{-1} \tag{2.6}$$

while the density of states $\rho(\delta)$ is given by the classical expression

$$\rho(\delta) = \frac{3V\delta_{ab}{}^2}{2\pi^2\hbar^3 v^3} \tag{2.7}$$

where the crystal has volume V and the velocity of sound in it is v. The difference in velocity in different directions of the crystal is neglected in this expression.

C

The square of the strain matrix element is given (see, for example, Abragam[17]) by

$$\mid \langle P_{nm} \mid \varepsilon \mid P_{nm} + 1 \rangle \mid^2 = \frac{(P_{nm}+1)\delta_{ab}}{2Mv^2}$$

where M is the mass of the crystal. Thus

$$W_{b \to a} = \frac{2\pi}{\hbar} \left[\frac{(P_{nm}+1)\delta_{ab}}{2Mv^2} \times \frac{3V\delta_{ab}^2}{2\pi\hbar^3v^3} \right] \mid \langle a \mid \sum_{n,m} V_n^m \mid b \rangle \mid^2$$

$$= \frac{3}{2\pi} \left(\frac{\delta_{ab}}{\hbar} \right)^3 \frac{P_{nm}+1}{\rho\hbar v^5} \, A^2 \; \text{sec}^{-1} \tag{2.8}$$

where ρ is the density of the crystal and we have written A^2 for $\mid \langle a \mid \sum V_n^m \mid b \rangle \mid^2$.

Similarly

$$W_{a \to b} = \frac{3}{2\pi} \left(\frac{\delta_{ab}}{\hbar} \right)^3 \frac{P_{nm}}{\rho\hbar v^5} \, A^2 \tag{2.9}$$

Following equation 1.14, we write

$$-\frac{dn_b}{dt} = n_b \, W_{b \to a} - n_a \, W_{a \to b}$$

[One can regard the summation in equation (2.5) as operating over all spin sites in the crystal in state $\mid b \rangle$, following Orbach. In this case, $W_{b \to a}$ is more correctly written as the number of spins making the transition per unit time. If this is done, the n_b and n_a occur in equations (2.8) and (2.9) on summation over j, the summation signs in these equations then only referring to n, m. The following equations agree with either approach.]

Thus

$$\frac{dn_b}{dt} = -\lambda[n_b(P_{nm}+1) - n_a P_{nm}] \tag{2.10}$$

where we have written

$$\lambda = \frac{3}{2\pi} \left(\frac{\delta_{ab}}{\hbar} \right)^3 \frac{A^2}{\rho\hbar v^5}$$

If all other levels in the spin system are unpopulated or their populations are independent of those of the two lowest states,

$$\frac{dn_b}{dt} = -\frac{dn_a}{dt} = \frac{1}{2} \frac{d(n_b - n_a)}{dt} = \frac{1}{2} \frac{dn}{dt}$$

Recalling that the thermal equilibrium values N_b and N_a are related by

$$N_a - N_b = N(\tanh \delta_{ab}/2kT) \qquad (1.18, \text{ p. } 12)$$

that $n_a + n_b = N_a + N_b = N$, and using equation (2.6) for P_{nm}, we can re-write equation (2.10) as

$$\frac{dn}{dt} = -2\pi \left[\frac{n}{\exp(\delta_{ab}/kT) - 1} + n_b \right]$$

$$= \frac{-2\lambda}{\exp(\delta_{ab}/kT) - 1} \left[n_b \exp(\delta_{ab}/kT) - n_a \right]$$

$$= \frac{\lambda \exp(\delta_{ab}/kT) + 1}{\exp(\delta_{ab}/kT) - 1} (n - n_0) \qquad (2.11)$$

From the definition in equation (1.21) we write the spin-lattice relaxation time for the direct process as

$$\frac{1}{T_1(\text{d})} = \lambda \coth \left(\frac{\delta_{ab}}{2kT} \right)$$

$$= \frac{3}{2\pi} \left(\frac{\delta_{ab}}{\hbar} \right)^3 \frac{\coth(\delta_{ab}/2kT)}{\rho \hbar v^5} \left| \langle a \, | \sum_{n,m} V_n{}^m \, | \, b \rangle \right|^2 \text{ sec}^{-1} \qquad (2.12)$$

When $\delta_{ab} \ll 2kT$, $\coth(\delta_{ab}/2kT) \to 2kT/\delta_{ab}$ and $1/T_1(\text{d})$ takes the form

$$\frac{1}{T_1(\text{d})} = \frac{3\delta_{ab}{}^2 kT}{\pi \hbar^4 v^5 \rho} \left| \langle a \, | \sum_{n,m} V_n{}^m \, | \, b \rangle \right|^2 \text{ sec}^{-1} \qquad (2.13)$$

A more rigorous treatment (see, for example, Stevens,[8] Orbach and Manenkov[18]) gives essentially the same expression as the above but the simplifications which have been introduced, such as the isotropic value of v, the assumption of a Debye lattice, and the assumption of only two populated spin levels, become more apparent.

In order to compute a value for T_1 we must know the appropriate parameters in the equation (2.12) as well as the square of the matrix elements shown. We shall consider this problem further below, and shall later compare theoretical and experimental values of T_1 in selected cases.

The difference in behaviour between Kramers and non-Kramers salts is shown by equation (2.12); in a Kramers salt, the matrix elements vanish since $| \, a \rangle$ and $| \, b \rangle$ are time conjugate states.

The direct process in non-Kramers salts

In these salts the magnetic ions have an even number of electrons and in general there exist non-zero matrix elements of $V_n{}^m$ connecting states $|b\rangle$ and $|a\rangle$. Thus equation (2.13) is obeyed in its present form and its temperature dependence is given by

$$\frac{1}{T_1(\mathrm{d})} = AT$$

Indeed, if it is valid to write $\delta_{ab} = g\beta H$ we have

$$\frac{1}{T_1(\mathrm{d})} = A'H^2T$$

To give a numerical example, we shall follow Orbach in calculating $1/T_1(\mathrm{d})$ for holmium ethyl sulphate.

Baker and Bleaney[15] studied the spin resonance spectra of this material and in the transition considered by Orbach, the relaxation occurs between

$$|b\rangle = \frac{1}{2}\Big[|6\rangle + |-6\rangle\Big] + \frac{1}{\sqrt{2}}|0\rangle$$

and a member of the ground doublet, for example

$$|a\rangle = 0.933\,|7\rangle + 0.342\,|1\rangle + 0.11\,|-5\rangle$$

(See the table on p. 21.)

Since $\mathscr{J} = 8$ in holmium, $V_n{}^1$ and $V_6{}^5$ have non-vanishing matrix elements between $|b\rangle$ and $|a\rangle$ where $n = 2$, 4 or 6.

Ignoring small terms we may write eventually

$$\langle a\,|\,\sum_{n,m} V_n{}^m\,|\,b\rangle = 3.9 \times 10^{-3} A_6{}^5 r^6 - 0.037\,A_2{}^1 r^2$$
$$- 0.046\,A_4{}^1 r^4 + 0.099\,A_6{}^1 r^6 \qquad (2.14)$$

Here, of course, the $A_n{}^m r^n$ are orbit-lattice, dynamic values. As the strain ε is small at temperatures much less than the melting point of the material, it can be used as an expansion parameter for the electrostatic potential energy in which the electrons of the magnetic ion move. As stated on p. 22, in this expansion the zero order term is the static $V_n{}^m$ while the first-order term gives the orbit-lattice interaction. As an approximation, found to work satisfactorily in calculation of phonon behaviour in dielectric crystals (Orbach[19]), the orbit lattice $A_n{}^m r^n$ is taken equal in magnitude to the static $A_n{}^m r^n$. This is a somewhat crude, certainly unjustified, assumption in the present circumstances, but it is the best that can be done at the moment.

Baker and Bleaney estimate as follows:

$$A_2{}^0 r^2 = 0, \quad A_4{}^0 r^4 = -18 \text{ cm}^{-1},$$

$$A_6{}^0 r^6 = 22 \text{ cm}^{-1}, \quad A_6{}^6 r^6 = 170 \text{ cm}^{-1}$$

Using the relations,

$$A_n{}^m r^n = A_n{}^0 r^n \text{ when } n = 2 \text{ or } 4 \text{ and}$$

$$A_6{}^m r^6 = [|\, A_6{}^0 r^6\,|^{\,6-|m|}\,|\, A_6{}^6 r^6\,|^{\,|m|}]^{1/6}$$

we find

$$A_6{}^5 r^6 = [(22)(170)^5]^{1/6} \simeq 120 \text{ cm}^{-1}$$

$$A_6{}^1 r^6 = [(22^5) 170]^{1/6} \simeq 30 \text{ cm}^{-1}$$

Thus the last term in equation (2.14) predominates and the matrix element is approximately equal to 3 cm^{-1}.

Using the rough values of $\rho = 2$ gm cm^{-3}, and $v = 2 \cdot 5 \times 10^5$ cm sec^{-1}, the spin-lattice relaxation time in zero field is given for this transition, where $\delta_{ab} = 5 \cdot 5$ cm^{-1}, by

$$T_1 = 2 \times 10^{-5} \tanh(3 \cdot 95/T) \text{ sec}$$

Thus a value of T_1 of the order of 20 μsec is predicted in the region of 1°K.

The direct process applied to Kramers salts

The equation (2.13) cannot be applied directly in the case of Kramers salts where the lowest states $|\, a \rangle$ and $|\, b \rangle$ are time conjugates of one another and are composed of states of half integral quantum numbers. Kramers' theorem is obeyed and the orbit-lattice interaction cannot connect these states. Thus in first order and in zero applied magnetic field, the matrix elements between $|\, a \rangle$ and $|\, b \rangle$ are identically zero. It is thus not possible to have direct relaxation between members of a Kramers pair, for with $H = 0$ the pair are clearly degenerate and lattice phonons of zero energy would be involved. A magnetic field can, however, split the pair and in doing so the states $|\, a \rangle$ and $|\, b \rangle$ are modified by the mixing in of some contributions from low lying excited states. A schematic diagram of the energy levels is given in Fig. 2.2.

If we consider just the first excited pair of states $|\, a_1 \rangle$ and $|\, b_1 \rangle$, the original states $|\, a \rangle$ and $|\, b \rangle$ become modified in an applied field $\mathbf{H}$ and take the form

$$|a'\rangle = |a\rangle - \frac{\beta\Lambda\mathbf{H}}{\Delta}\cdot\langle a_1 | \mathbf{J} | a\rangle | a_1\rangle$$

$$|b'\rangle = |b\rangle - \frac{\beta\Lambda\mathbf{H}}{\Delta}\cdot\langle b_1 | \mathbf{J} | b\rangle | b_1\rangle$$

where Λ is the Landé splitting factor.

Thus the time conjugate nature of the pair of states is removed and matrix elements of the form

$$\langle a' | \sum_{n,m} V_n{}^m | b'\rangle$$

can be found which have non-zero values.

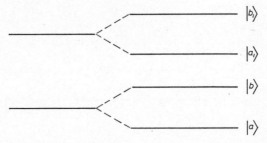

F IG. 2.2. Hypothetical level diagram for a Kramers salt, as considered in the text. In zero magnetic field (left-hand side of diagram) each level is doubly degenerate, the degeneracy being removed by an applied field (right-hand side of diagram). States $|a\rangle | b\rangle$ and $|a_1\rangle | b_1\rangle$ are of the form $|+\tfrac{1}{2}q\rangle | -\tfrac{1}{2}q\rangle$, where q is an odd integer.

Pursuing this calculation, Orbach has shown that

$$\frac{1}{T_1(\mathrm{d})} = \frac{3\,\delta_{ab}{}^3 4\beta^2\Lambda^2\,|\,\mathbf{H}.\,\langle a_1 | \mathbf{J} | a\rangle\,|^2\,|\,\langle a | \sum_{n,m} V_n{}^m | b\rangle\,|^2}{2\pi\rho v^5\hbar\,\Delta_c{}^2\,\tanh(\delta_{ab}/2kT)} \tag{2.15}$$

If we define a unit vector $\mathbf{h}$ in the direction of the external field and let the splitting $\delta_{ab} = g\beta H$, where $g^2 = g_{\parallel}{}^2\cos^2\theta + g_{\perp}{}^2\sin^2\theta$, $g_{\parallel}$ and $g_{\perp}$ being those values appropriate to the ground doublet, we get for $g\beta H < kT$

$$\frac{1}{T_1(\mathrm{d})} = \frac{12\beta^4 g^2\Lambda^2 H^4 kT\,|\,\mathbf{h}.\,\langle a_1 | \mathbf{J} | a\rangle\,|^2\,|\,\langle a | \sum_{n,m} V_n{}^m | b\rangle\,|^2}{\pi\rho v^5\hbar^4\Delta_c{}^2} \tag{2.16}$$

Thus for Kramers salts we have the condition

$$\frac{1}{T_1(\mathrm{d})} \propto H^4 T$$

The calculation of $1/T_1$ based upon equation (2.16) is given by

Orbach for the case of dysprosium ethyl sulphate. Using experimentally determined values for the various parameters, he finds

$$\frac{1}{T_1(\text{d})} \approx 1 \cdot 7 \times 10^{-11} H^4 T \sin^2 \theta \cos^2 \theta$$

when the external magnetic field makes an angle θ with the crystal z axis. $T_1(\text{d})$ is thus estimated to be about 3 msec in fields of 3000 gauss at $4 \cdot 2°\text{K}$, with $\theta = 45°$.

Clearly this equation does not apply when $\theta = 0$. The levels a_1, b_1 must be re-chosen so that a finite transition probability occurs when $\theta = 0$. In fact, for dysprosium ethyl sulphate, the first such excited level is 59 cm^{-1} above the ground level and the matrix element is found to be 250 times smaller than that which obtains when $\theta \neq 0$. The calculation yields a relaxation time 6000 times longer than that at $\theta = 45°$ and probably explains why the direct process has not been observed experimentally in this material (Gill[20]). The Orbach two-phonon process (see below) was apparently much more rapid, even at $2°\text{K}$.

An example of the use of these equations for $1/T_1(\text{d})$ comparing the predicted results with an experimental determination of relaxation rates is afforded by the work of Scott and Jeffries. We shall give details of this later.

The two-phonon relaxation process

In considering the direct one-phonon process, we assumed that a spin flip between states $|b\rangle$ and $|a\rangle$ was accompanied by the creation or annihilation of a single phonon. The orbit-lattice interaction was used once only to effect this, and the single phonon had energy equal to the splitting between states $|b\rangle$ and $|a\rangle$. The two-phonon process is of the next higher order in time-dependent perturbation theory, in that V_{ol} is involved twice. There results the creation and destruction of a pair of phonons and their *difference* in energy is equal to the splitting between $|b\rangle$ and $|a\rangle$. [The terms in ε^2 of equation (2.3) contribute a first-order perturbation but they are usually small. See, for example, Stoneham.[21]]

The state of affairs is shown diagrammatically in Fig. 2.3. Relaxation from state $|b\rangle$ to state $|a\rangle$ proceeds through a state $|c\rangle$ and phonon modes designated nm and $n'm'$ are involved. We assume there are P_{nm} and $P_{n'm'}$ phonons in those two modes before the transitions occurs and that these change to $P_{nm}+1$, $P_{n'm'}-1$ when the transition is complete.

The energy balance is satisfied if the following equation is obeyed.

$$\delta_{ab} = \hbar\omega_{nm} - \hbar\omega_{n'm'}$$

Thus we can think of the phonon $\hbar\omega_{n'm'}$ being absorbed by the spin system which is then excited to $|c\rangle$ followed by a transition to state $|a\rangle$ with emission of a phonon of energy $\hbar\omega_{nm}$. Our calculation of $1/T_1(\mathrm{d})$ involved matrix elements of the type $\langle a\,|\,\sum V_n{}^m\,|\,b\rangle$. Now a similar argument must be applied which will take us through $|c\rangle$ and we shall, in second order, get matrix elements connecting this level to $|a\rangle$ and $|b\rangle$, namely terms of the form

$$\langle a\,|\,\sum V_n{}^m\,|\,c\rangle\,\langle c\,|\,\sum V_{n'}{}^{m'}\,|\,a\rangle$$

The calculation involved has been followed by a number of authors and the reader is again referred to Orbach for details.

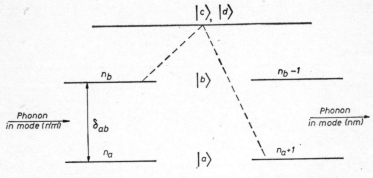

FIG. 2.3. Diagram of the two-phonon process. A transition from $|b\rangle$ to $|c\rangle$ occurs with absorption of a phonon of energy $\hbar\omega_{n'm'}$; a transition from $|c\rangle$ to $|a\rangle$ completes the relaxation process, with emission of a phonon of energy $\hbar\omega_{nm}$. $\hbar\omega_{nm}-\hbar\omega_{n'm'}=\delta_{ab}$.

Following precisely the arguments used for $T_1(\mathrm{d})$ Orbach finds for the two-phonon case, assuming $\delta_{ab}\ll kT$ or Δ_c

$$\frac{1}{T_1}=9/(16\pi^3\rho^2 v^{10})\cosh(\delta_{ab}/2kT)\quad\times$$

$$\int\left|\sum_{\substack{n,m,\\n',m',}}\frac{\langle a\,|\,V_n{}^m\,|\,c\rangle\,\langle c\,|\,V_{n'}{}^{m'}\,|\,b\rangle}{\hbar\omega_{n'm'}-\Delta_c}\right|^2\quad\times$$

$$\mathrm{cosech}^2\left(\frac{\hbar\omega_{n'm'}}{2kT}\right)\,\omega_{n'm'}^6\,d\omega_{n'm'}$$

$$(2.17)$$

The Raman process, $k\theta_D < \Delta_c$, applied to non-Kramers salts

As indicated in Chapter 1, the Raman process merely requires the energy balance indicated in Fig. 2.3, without c having a real existence. It is a virtual level.

Mathematically $\hbar\omega_{n'm'}$ can never equal Δ_c and the integrand in equation (2.17) is therefore peaked about $\hbar\omega_{n'm'} = kT$. Note that the upper limit to the integral is the limiting Debye frequency $\omega_{n'm'}$.

The relaxation time then becomes to a good approximation

$$\frac{1}{T_1(\mathrm{R})} = \frac{9(6!)}{4\pi^3\rho^2 v^{10}\Delta_c^2}\left(\frac{kT}{\hbar}\right)^7 \left| \sum_{\substack{n,m \\ n',m'}} \langle a | V_n^m | c \rangle \langle c | V_{n'}^{m'} | b \rangle \right|^2$$

(2.18)

This is the Raman process result first predicted by Kronig.

The efficiency of the Raman process becomes apparent now. The number of phonons present in the lattice with frequency between ω and $\omega + d\omega$ is

$$\frac{3V\omega^2 d\omega}{8\pi^2 v^3 [\exp(\hbar\omega/kT) - 1]}$$

so that as ω becomes small so does this number. In the one-phonon process, $\delta_{ab} = \hbar\omega$ is very much less than kT at even moderate temperatures so very little of the phonon spectrum is used. However, the phonons are more numerous in the two-phonon process, since the frequencies are very much higher, and many different pairs of frequencies can be utilized, since only their difference is determined. The matrix elements appearing in the higher-order transition are themselves much smaller than those in the one-phonon process, but the phonons involved are more numerous, thus offsetting the decrease to a considerable extent as the temperature is raised.

Orbach notes that for $\delta_{ab} \sim 1 \text{ cm}^{-1}$ and $\Delta_c \sim \langle a | V_n^m | c \rangle$, $T_1(\mathrm{d})/T_1(\mathrm{R}) \sim 10^{-9}T^6$. Thus at temperatures above about 30°K the Raman process dominates over the direct process, while at low temperatures the direct process normally takes complete control.

The Raman process, $k\theta_D < \Delta_c$, applied to Kramers salts

We now have time conjugate states $| a \rangle$ and $| b \rangle$ separated by an energy Δ_c from another pair of time conjugate states $| a_1 \rangle$ and $| b_1 \rangle$, as shown in Fig. 2.2. These excited states are thus available to act in the same way as state $| c \rangle$ in the above analysis, but with the time conjugate limitations. We shall not give details of the analysis which may be again found in the references cited above. One is finally led to the following expression for the relaxation time

$$\frac{1}{T_1(\text{R})} = \frac{9!\hbar^2}{\pi^3\rho^2v^{10}\Delta_c^4}\left(\frac{kT}{\hbar}\right)^9 |\sum_{\substack{n,m \\ n',m'}} \langle a \mid V_n{}^m \mid b_1 \rangle \langle b_1 \mid V_{n'}{}^{m'} \mid b \rangle |^2$$

(2.19a)

If the system is similar to that which was analysed in dealing with the direct process for Kramers salts, we must take into account the admixture of terms through the application of the magnetic field. This produces a Raman-like dependence of $T_1 \propto H^{-2}T^{-7}$ as well as T^{-9}. The dominant term will generally be that in T^{-9} but this need not be so, if there are significant admixtures into states $\mid a_1 \rangle$ and $\mid b_1 \rangle$ of further nearby excited states, these admixtures being caused by the magnetic field.

We may thus summarize the state of affairs for the Raman relaxation by writing

Non-Kramers salts $\quad\quad \dfrac{1}{T_1(\text{R})} = BT^7$

Kramers salts $\quad\quad \dfrac{1}{T_1(\text{R})} = B'T^9 + B''H^2T^7$

The Orbach or resonant two-phonon process

In the Raman process there were no phonons of energy Δ_c, so that transitions involving a virtual level c had to be imagined. We now consider the state of affairs when the first excited level c (of Fig. 2.3) is sufficiently close to the ground levels a and b so that $h\theta_D > \Delta_c$. This means that the first excited level has an energy separation from the ground level less than the limiting Debye frequency.

Equation (2.17) still describes the situation, but the integrand is now different in form in that it peaks when $\hbar\omega_{n'm'} = \Delta_c$. (At first sight one would believe the integrand then to become infinite; this is not so, since the excited state $\mid c \rangle$ has a finite lifetime so that its energy is complex: this avoids the divergence.)

Orbach deduces for this relaxation rate

$$\frac{1}{T_1(\text{O})} = \frac{3}{2\pi\rho v^5\hbar^5}\left(\frac{\Delta_c}{\hbar}\right)^3 \left(\exp\frac{\Delta_c}{kT} - 1\right)^{-1}$$

$$|\sum_{\substack{n,m \\ n',m'}} \langle a \mid V_n{}^m \mid c \rangle \langle c \mid V_{n'}{}^{m'} \mid b \rangle |^2$$

$$\times \frac{}{|\sum_{n,m} \langle a \mid \overline{V_n{}^m \mid c} \rangle |^2 + |\sum_{n',m'} \langle c \mid V_{n'}{}^{m'} \mid b \rangle |^2} \quad\quad (2.19b)$$

In practice we generally have $\Delta_c \gg kT$ so that $[\exp(\Delta_c/kT) - 1]^{-1}$ just becomes $\exp(-\Delta_c/kT)$. Hence

$$\frac{1}{T_1(0)} = C \exp(-\Delta_c/kT)$$

The equation holds for both non-Kramers and Kramers salts provided we can consider $|c\rangle$, $|d\rangle$ as equivalent to the states $|a_1\rangle$, $|b_1\rangle$, the first excited states of the earlier Raman analysis.

§2.2

THE VAN VLECK-MATTUCK AND STRANDBERG APPROACH

In this section we shall outline the method of calculation of T_1 developed first by Van Vleck[6,7] and later by Mattuck and Strandberg[22] and others. It is worth repeating, however, that the difference from the work of §2.1 lies in the mathematics of the method and not in the physics of the model. In each case, an expression is found for the orbit-lattice interaction, V_{ol}, which, combined with the spin-orbit coupling, causes the spin-lattice relaxation; then the transition probability is calculated, finding the matrix elements of V_{ol} between appropriate eigenstates of the spin and lattice Hamiltonians, which we write as $\mathcal{H}_{spin}$ and $\mathcal{H}_{lattice}$ respectively.

We follow the treatment of Mattuck and Strandberg and note first the assumptions of their treatment: (*a*) All effects of spin–spin interaction are neglected, except those leading to energy-level broadening. (*b*) The effects of local lattice distortion (strictly, the effects on lattice modes) produced by the presence of the paramagnetic ion are ignored. (*c*) The crystalline electric field produces a potential energy term which is smaller than the potential energy of the free ion, but greater than the spin-orbit energy. This implies that the results will be applicable to iron group ions and not to ions of the rare earths. (*d*) The theory is also not applicable to ions whose ground state is $L = 0$.

The total Hamiltonian of the paramagnetic ion-lattice system in a magnetic field H is written as

$$\mathcal{H} = \mathcal{H}_L + \mathcal{H}_0 + V + 2\beta \mathbf{S}.\mathbf{H} + \lambda \mathbf{L}.\mathbf{S} + \beta \mathbf{L}.\mathbf{H} \qquad (2.20)$$

Here λ is the spin-orbit coupling parameter, V the potential energy of the ion due to the crystalline electric field, $\mathcal{H}_0$ the free-ion energy and $\mathcal{H}_L$ is the energy of the lattice. The latter is written as

$$\mathcal{H}_L = \sum_{nm} \hbar\omega_{nm}(a_{nm}{}^\dagger a_{nm} + \tfrac{1}{2})$$

The index nm represents the phonon mode and branch number, while ω_{nm} is the corresponding frequency. The annihilation and creation operators, $a_{nm}\dagger$ and a_{nm} respectively, have matrix elements

$$\langle P_{nm} \mid a_{nm} \mid P_{nm} + 1 \rangle = \langle P_{nm} + 1 \mid a_{nm}\dagger \mid P_{nm} \rangle = (P_{nm} + 1)^{\frac{1}{2}}$$

where P_{nm} is the phonon occupation number of the mode nm in thermal equilibrium and

$$P_{nm} = [\exp(\hbar\omega_{nm}/kT) - 1]^{-1}$$

as in equation (2.6).

We now consider an atom in its equilibrium position, $\mathbf{r}$, and allow a displacement to occur. We write $u_{r\alpha}(\alpha = x, y, z)$ for this displacement and, following Klemens[23]

$$u_{r\alpha} = \left(\frac{2\hbar}{M}\right)^{\frac{1}{2}} \sum_{nm} \frac{1}{(\omega_{nm})^{\frac{1}{2}}} \phi_{nm\alpha} (a_{nm} + a_{nm}\dagger) \cos(\mathbf{k}_{nm}.\mathbf{r} + \Delta_{nm})$$

$$(2.21)$$

where M is the mass of the crystal, $\phi_{nm\alpha}$ is the αth component of the unit polarization vector for the mode nm and $\mathbf{k}_{nm}$ is its propagation vector. Δ_{nm} is an arbitrary phase factor. For simplicity, it is assumed that all phonons have the same velocity v and the density of states is therefore given by the Debye formula, equation (2.7), where we may write $\hbar\omega_{nm} = \delta_{ab}$.

As in §2.1, we require an expression for the variation of the crystalline potential V, due to relative displacements of an ion and its neighbours. V is expanded in a power series in the normal displacements, Q_f, of the nearest neighbours (Van Vleck[6])

$$V = V_0 + \sum_f \frac{\partial V}{\partial Q_f} Q_f + \frac{1}{2} \sum_{ff'} \frac{\partial^2 V}{\partial Q_f \partial Q_{f'}} Q_f Q_{f'} + \ldots \quad (2.22)$$

[Equation (2.22) is equivalent to equation (2.4) of the previous analysis.]

The Q's are related to the ordinary displacements of the neighbours $(\delta R_{l\alpha})$ by the equation

$$Q_f = \sum_{l\alpha} B_{fl\alpha} \, \delta R_{l\alpha} \quad (2.23)$$

The $\delta R_{l\alpha}$ may be expanded in normal lattice modes by equation (2.21). Assuming $\mathbf{k}_{nm}.\mathbf{r}$ is very small, and ignoring the resulting term in (2.21) which is independent of **r**, since only relative displacements will modulate the crystal field, we have

$$\delta R_{l\alpha} = \left(\frac{2\hbar}{M}\right)^{\frac{1}{2}} \sum_{nm} \left(\frac{\omega_{nm}}{v}\right) \; \phi_{nm\alpha}(a_{nm} + a_{nm}\dagger) \; \mathbf{K}_{nm}.\mathbf{R}_l \sin \Delta_{nm}$$

$$(2.24)$$

where $\mathbf{K}_{nm}$ is a unit vector in the $\mathbf{k}_{nm}$ direction and $|k_{nm}| = \omega_{nm}/v$. From equations (2.22), (2.23) and (2.24) we can write, following Mattuck and Strandberg,

$$V = V_0 + \sum_{fnm} V^f A^{fnm} \Pi_{nm} + \sum_{\substack{fnm \\ f'n'm'}} V^{ff'} A^{fnm} A^{f'n'm'} \Pi_{nm} \Pi_{n'm'} + \cdots$$

$$(2.25)$$

where we have written

$$V^f = \frac{\partial V}{\partial Q_f} \; ; \quad V^{ff'} = \frac{1}{2} \frac{\partial^2 V}{\partial Q_f \partial Q_{f'}} \; ; \quad \Pi_{nm} = (a_{nm} + a_{nm}\dagger)$$

and

$$A^{fnm} = \left(\frac{2\hbar\omega_{nm}}{Mv^2}\right)^{\frac{1}{2}} \sin \Delta_{nm} \sum_{l\alpha} B_{fl\alpha} \mathbf{K}_{nm}.\mathbf{R}_l$$

Thus the total Hamiltonian, equation (2.20), may be written

$$\mathcal{H} = \sum_{nm} \hbar\omega_{nm}(a_{nm}\dagger a_{nm} + \tfrac{1}{2}) + \mathcal{H}_0 + V_0 + 2\beta\mathbf{S}.\mathbf{H} + \lambda\mathbf{L}.\mathbf{S}$$

$$+ \beta\mathbf{L}.\mathbf{H} + \sum_{fnm} V^f A^{fnm} \Pi_{nm}$$

$$+ \sum_{\substack{fnm \\ f'n'm'}} V^{ff'} A^{fnm} A^{f'n'm'} \Pi_{nm} \Pi_{n'm'}$$

$$(2.26)$$

The analysis so far is common to both the original treatment of Van Vleck and the later paper of Mattuck and Strandberg; both proceed to divide $\mathcal{H}$ into $\mathcal{H}_{\text{lattice}}$, $\mathcal{H}_{\text{spin}}$ and $\mathcal{H}_{\text{interaction}}$, the last being that part of the Hamiltonian which induces the spin-lattice relaxation. Mattuck and Strandberg write

$$\mathcal{H}_{\text{lattice}} = \sum_{nm} \hbar\omega_{nm}(a_{nm}\dagger a_{nm} + \tfrac{1}{2}) \tag{2.27}$$

$$\mathcal{H}_{\text{spin}} = \mathcal{H}_0 + V_0 + 2\beta\mathbf{S}.\mathbf{H} + \lambda\mathbf{L}.\mathbf{S} + \beta\mathbf{L}.\mathbf{H} \tag{2.28}$$

$$\mathcal{H}_{\text{interaction}} = \sum_{fnm} V^f A^{fnm} \Pi_{nm} + \sum_{\substack{fnm \\ f'n'm'}} V^{ff'} A^{fnm} A^{f'n'm'} \Pi_{nm} \Pi_{n'm'}$$

$$(2.29)$$

Van Vleck includes the last two terms on the right-hand side of (2.28) in his expression for $\mathcal{H}_{\text{interaction}}$, and omits them from his

$\mathcal{H}_{spin}$. The above formulation is to be preferred since the $\mathcal{H}_{spin}$ of (2.28) is precisely that which leads to the spin Hamiltonian describing the energy levels observed in laboratory E.S.R. experiments (see, for example, Assenheim[24], Low[25]).

We are indebted to Dr Stoneham for the summary of the several approaches to the calculation which is given in Table 2.2.

TABLE 2.2. A comparison of treatments of the relaxation calculation.

The total Hamilton $\mathcal{H}$ is written as

$$\mathcal{H} = \mathcal{H}_{lattice} + \mathcal{H}_{spin} + \mathcal{H}_{interaction}$$

$$= \underbrace{\mathcal{H}_L + \mathcal{H}_0 + V_0} + \underbrace{2\beta S.H + \lambda L.S + \beta L.H} + \mathcal{H}_{interaction}$$
$$\quad\quad \mathcal{H}_A \quad\quad\quad\quad G$$

Authors	Treat exactly	Löwdin's[26] method expansion	Treat as perturbation
Mattuck and Strandberg[22] Stoneham[21]	$\mathcal{H}_L + \mathcal{H}_A$	G	$\mathcal{H}_{interaction}$
Van Vleck[7]	$\mathcal{H}_L + \mathcal{H}_A$		$G + \mathcal{H}_{interaction}$
(a) Foglio[27] (b) Stoneham[28]	$\mathcal{H}_L + \mathcal{H}_A$ + spin-orbit coupling within lowest states		$\beta L.H$ + $\mathcal{H}_{interaction}$
(c) Orbach[29]	$\mathcal{H}_L' + \mathcal{H}_A + G$		$\mathcal{H}_{interaction}$

(a) Refers to the Co^{2+} ion in MgO. (b) Refers to the Co^{2+} ion in Al_2O_3. (c) Formal treatment, no case explicitly considered.

The calculation now involves diagonalizing $\mathcal{H}_{spin}$ to an appropriate order—probably second order—finding the energy level E_i and the corresponding state vectors ψ_i and then calculating the required matrix elements of $\mathcal{H}_{interaction}$ between simultaneous eigenstates of $\mathcal{H}_{spin}$ and $\mathcal{H}_{lattice}$. We shall not follow this calculation through; instead the reader is referred to the original papers already cited and to the work of Stoneham.[21]

As we have seen, the direct relaxation process involves a spin transition to a lower energy state and the simultaneous creation of a phonon so that a direct energy balance occurs (Figs. 2.1 and 2.2). Mattuck and Strandberg calculate $\mathcal{H}_{direct}$, which gives the spin-phonon interaction between the two states $|a\rangle$ and $|b\rangle$ of the spin Hamiltonian. They then calculate the transition probability from

$$W = \left(\frac{1}{\hbar^2}\right) \mid \langle P_{nm}, b \mid \mathscr{H}_{direct} \mid P_{nm} + 1, a \rangle \mid^2 \rho(\omega_r) \qquad (2.30)$$

where ω_r is the resonance frequency. This equation is equivalent to (2.5) of the previous section.

An order of magnitude estimate of the parameters involved results in the following very approximate expression for W when the direct process occurs:

$$W \approx \frac{3456\pi^2 k\omega_r^2}{\hbar^2(M/V)v^5} \left(\frac{e^4 r_0^4}{\Delta_c^4 R^6}\right) \mid 2\lambda\beta HS + \lambda^2 S_A \mid_{a,b}^2 \qquad (2.31)$$

where r_0 is the ionic radius, R the average ionic separation and S_A the spin anticommutator. The other parameters have the usual significance, noted earlier in this chapter.

Riggs[30] measured the spin-lattice relaxation of the Cu^{2+} ion in the zinc tungstate lattice, and found values for T_1 at $4 \cdot 2°K$ of the order of $1-5$ msec. Equation (2.31) predicts a similar order of magnitude when plausible values for the various parameters are inserted.

Stoneham[21] used this approach in a detailed and more exact examination of the relaxation of the Cu^{2+} ion substituted for zinc in the potassium zinc Tutton salt lattice. It is significant that *both the magnitude and the anisotropy* of the calculated relaxation rate agreed well with the experimental results of Gill.[31]

§2.3
SUMMARY OF THE ANALYSES

We have presented only the outline of the theoretical analyses of the basic spin relaxation phenomena in the preceding pages. We may summarize the findings in the following way.

We assume a perfect crystal of which the lattice temperature is always identical with the temperature of the bath surrounding the crystal. Inherent in the analysis is the assumption that the ground level in the case of the non-Kramers salt or the ground doublet in the case of the Kramers salt remains always well separated from the excited states.

Within these limits then we find

(a) for a non-Kramers salt

$$\frac{1}{T_1} = AT + BT^7 + C \exp(-\Delta_c/kT) \qquad (2.32)$$

In certain cases we may replace A by $A''H^2$.

(b) for a Kramers salt

$$\frac{1}{T_1} = A'H^4T + B''H^2T^7 + B'T^9 + C' \exp(-\Delta_c/kT) \quad (2.33)$$

where A, B, C, A', B', B'', C' are constants and Δ_c is the energy separation of the relevant excited level.

Orbach and Blume[32] considered the case where a multi-level Kramers state is lowest in energy and concluded that a state of affairs may exist where a T^5 term dominates the relaxation rate, in at least the lower region of the Raman temperature range. They showed that an order of magnitude criterion for the possible dominance of the T^5 term over the T^7 term is $\lambda^2/\Delta > kT$ where λ is the spin-orbit coupling constant and Δ is the appropriate crystal-field splitting. This criterion is applicable to ions of the first two transition groups. In the case of the rare-earth ions, where the crystal-field splittings are small compared with the spin-orbit splittings, the criterion becomes simply $\Delta > kT$.

There is some experimental evidence for the existence of a T^5 law in the measurements of Castle and Feldman.[33] These relate to the divalent vanadium ion in MgO. In the temperature region 16–90°K the experimental results indicate a T^5 law, but this range of temperature is considerably larger than the theoretical analysis would lead one to suppose possible. Moreover, this temperature dependence is not found in the case of the isoelectronic Cr^{3+} ion in MgO, for which λ/Δ has about the same value. Huang[34] found a T^5 temperature dependence of the relaxation of the Eu^{2+} ion in CaF_2 throughout the temperature range 15–30°K. Bierig *et al.*[35] report similar results for Gd^{3+} in CaF_2, while Horak and Nolle[36] found this law to hold for Eu^{2+} in BaF_2 and for Mn^{2+} in BaF_2 and SrF_2.

If the T^5 law exists in other cases but extends over a small temperature region, it may well be difficult to establish experimentally with any great certainty since it is predicted to occur in just that region where direct and Raman (including Orbach's) terms have similar orders of magnitude and a clear simple dependence on temperature does not then occur.

In many cases of interest the analysis outlined in this chapter will not apply in its present form since it does not describe precisely the experimental conditions. For example, the trivalent chromium ion has been widely studied in ruby and we shall deal with this later in Chapter 9. The Cr^{3+} ion in zero field has the $\pm\frac{3}{2}$ levels lowest, but only about 0·4 cm^{-1} below the $\pm\frac{1}{2}$ spin levels. When a field is applied and the crystal orientation is varied, the four spin levels alter in separation and in character in a marked manner. Understandably in such a case the theoretical expressions of this chapter are not adequate although the same form of the temperature dependence is found experimentally. In particular, difficulty is encountered in defining a spin temperature, equation (1.11) referring to only two levels. In thermal equilibrium all spin temperatures defined for any pair of levels in the multi-level system are equal to one another and to the lattice temperature. If, however, the populations of any two levels are disturbed, this uniqueness is lost, and the concept of a spin temperature, which applies even in non-equilibrium conditions for a two-level system, can now only be applied with great care.

Another point is that all populated levels may now take part in the recovery to thermal equilibrium of any pair of levels. Thus the rate equations for a multi-level system may involve several time constants, none of which can be related in a simple manner to the relaxation time $T_1(i, j)$ of a pair of levels i, j. An example of this is to be found in the calculations of Donoho,[37] who has attempted to solve the rate equations for the four-level Cr^{3+} system mentioned above. This will be discussed in more detail in Chapters 7 and 9.

References

1. Waller, I., *Z. Physik*, 1932, **79**, 370.
2. Gorter, C. J., *Paramagnetic Relaxation* (Elsevier Publishing Co. Inc., 1947).
3. Heitler, W., and Teller, E., *Proc. Roy. Soc.*, 1936, **155**A, 629.
4. Fierz, M., *Physica*, 1938, **5**, 433.
5. Kronig, R. de L., *Physica,* 1939, **6**, 33.
6. Van Vleck, J. H., *J. Chem. Phys.*, 1939, **7**, 72.
7. Van Vleck, J. H., *Phys. Rev.*, 1940, **57**, 426.
8. Stevens, K. W. H., *Repts. on Prog. in Phys.*, 1967, **30**, 189.
9. Scott, P. L., and Jeffries, C. D., *Phys. Rev.*, 1962, **127**, 32.
10. Hutchings, M. T., *Solid State Physics*, Vol. 16 (Academic Press, 1964).
11. Stevens, K. W. H., *Proc. Phys. Soc.*, 1952, **65**A, 209.
12. Elliott, R. J., and Stevens, K. W. H., *Proc. Roy. Soc.*, 1952, **215**A, 437.
13. Orbach, R., *Proc. Roy. Soc.*, 1961, **264**A, 458.
14. Elliott, R. J., and Stevens, K. W. H., *Proc. Roy. Soc.*, 1953, **218**A, 553.
15. Baker, J. M., and Bleaney, B., *Proc. Roy. Soc.*, 1958, **245**A, 156.
16. Schiff, L. I., *Quantum Mechanics*, 2nd Ed. (McGraw Hill, 1955).
17. Abragam, A., *Principles of Nuclear Magnetism* (O.U.P., 1961).
18. Manenkov, A. A., and Orbach, R., *Spin-Lattice Relaxation in Ionic Solids* (Harper and Row, 1966).
19. Orbach, R., Ph.D. thesis, Berkeley, University of California, 1960.
20. Gill, J. C., *Proc. Phys. Soc.*, 1963, **82**, 1066.
21. Stoneham, A. M., *Proc. Phys. Soc.*, 1965, **85**, 107.
22. Mattuck, R. D., and Strandberg, M. W. P., *Phys. Rev.*, 1960, **119**, 1204.
23. Klemens, P. G., *Solid-State Physics*, Vol. 7 (Academic Press, New York, 1958).

24. Assenheim, H. M., *Introduction to Electron Spin Resonance* (Adam Hilger Ltd, 1966; Plenum Press, New York, 1967).
25. Low, W., *Paramagnetic Resonance in Solids* (Academic Press, 1960).
26. Löwdin, P. O., *J. Chem. Phys.*, 1951, **19**, 1396.
27. Foglio, M. E., Ph.D. thesis, Bristol University, 1962.
28. Stoneham, A. M., Ph.D. thesis, Bristol University, 1964.
29. Orbach, R., *Proc. Phys. Soc.*, 1961, **77**A, 821.
30. Riggs, R. J., Ph.D. thesis, St Andrews University, 1968.
31. Gill, J. C., *Proc. Phys. Soc.*, 1965, **85**, 119.
32. Orbach, R., and Blume, M., *Phys. Rev. Letters*, 1962, **8**, 478.
33. Castle, J. G., and Feldman, D. W., quoted in reference 27.
34. Huang, Chao-Yuan, *Phys. Rev.*, 1965, A**139**, 241.
35. Bierig, R. W., Weber, M. J., and Warshaw, S. I., *Phys. Rev.*, 1964, **134**, A1504.
36. Horak, J. B. and Nolle, A. W., *Phys. Rev.*, 1967, **153**, 372.
37. Donoho, P. L., *Phys. Rev.*, 1964, A**133**, 1080.

Chapter 3

Cross Relaxation

When discussing spin-lattice relaxation mechanisms in the previous chapter, it was assumed that two spin levels only were involved in the relaxation phenomena. In the case of the direct process in Kramers salts it was necessary to invoke the presence of higher energy levels, but their effect was confined to the mixing of some small parts of their wave functions, through the application of the steady magnetic field. Thus we had assumed an effective spin $S = \frac{1}{2}$ for the magnetic ions. In the present chapter we shall not consider the general case of $S > \frac{1}{2}$; we postulate instead the presence of more than two spin levels, but we require a simple relation between the splittings of at least two pairs of the levels, in order that *cross relaxation* can occur.

The first theoretical and experimental discussion of the phenomenon was in the celebrated paper by Bloembergen, Shapiro, Pershan and Artman[1] in 1959. Indeed, the name 'cross relaxation' probably first appeared in that paper. In effect, Bloembergen *et al.* considered the energy transfer which might occur when one has a simple system such as that shown in Fig. 3.1. Two levels, 2 and 3, are close together and separated from level 1. The frequencies v_{12} and v_{13} are thus similar but they are not identical, although the difference $(v_{13} - v_{12})$ is very much less than v_{12} or v_{13}. In the case where the absorption lines for the 1–2 and 1–3 transitions overlap, albeit only in the wings, Bloembergen *et al.* showed that there is a finite probability of a transition down, say, by a spin in level 3 to level 1 accompanied by the *simultaneous* transition up of another spin from level 1 to level 2, the small unbalance of energy being taken up by the dipolar or internal energy of the spin system. If we call this transition probability per spin w, it is possible to define the cross-relaxation time T_{21} as $T_{21} = 1/(2w)$. (It is interesting to note here a difference in custom amongst different writers. Bloembergen *et al.* designated the cross-relaxation time as T_{21} to indicate its intermediate position between the spin-lattice relaxation time T_1 and the spin-spin relaxation time T_2. Some authors have used the nomenclature T_{ij}, where i and j are the levels between which cross relaxation occurs; thus in Fig. 3.1 instead of writing T_{21} these authors would write T_{23}. We shall use the Bloembergen nomenclature.)

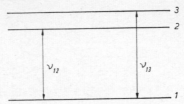

FIG. 3.1. Simple energy-level diagram wherein spin transitions $3 \to 1$ and $1 \to 2$ nearly conserve energy.

The physical nature of the process

Consider a spin system with several spin levels, at a uniform temperature T. The populations of these levels will be given by a Boltzmann distribution. Suppose now that this distribution is disturbed and the disturbing influence is then withdrawn. We can distinguish clearly two ways in which the Boltzmann distribution is regained. (*a*) If the energy levels are equally spaced, the various spins readily communicate with one another through the system and equilibrium is restored in a time of order T_2, the spin–spin relaxation time. (*b*) If the energy levels are unequally spaced, this means of communication directly from spin to spin is no longer possible and the equilibrium of each spin species is established directly with the lattice in a relaxation time T_1, the spin-lattice relaxation time. We are now interested in a third process, where the spin levels are not uniformly spaced but are only just not so. As indicated above, this process is again a spin–spin type, involving neighbouring spins wherein a transition down at spin A and a simultaneous transition up at spin B nearly conserve Zeeman energy, the balance being taken up by the internal energy of the spin system.

Transition probabilities and the rate equations

The transition probability of the third process which we have just described may be written as

$$w_{AB} = \hbar^{-2} \left| \langle (E_A, E_B) | \mathscr{H}_{AB} | E_A + h\nu_{12}, E_B - h\nu_{13} \rangle \right|^2 g(\nu)_{12,13}$$

(3.1)

Here E_A and E_B are the energies of the A and B spin systems respectively, $\mathscr{H}_{AB}$ is the interaction between two spins, one of type A and the other of type B, and $g(\nu)_{12, 13}$ is the shape function describing the process. In terms of the shape functions of the individual lines due to transitions of the ions A and B, $g(v)_{12}$ and $g(v)_{13}$, we may write the transition probability approximately as

$$w_{AB} = \hbar^{-2} \, | \, \mathcal{H}_{AB} \, |^2 \iint g(v')_{12} g(v'')_{13} \delta(v'-v') dv' dv'' \quad (3.2)$$

where the integrations represent the overlap between the two resonances. Bloembergen *et al.* showed that when the two magnetic resonances are Gaussian in shape, the double integral becomes

$$(2\pi)^{-\frac{1}{2}} [(\Delta v_{12})^2 + (\Delta v_{13})^2]^{-\frac{1}{2}}$$
$$\times \exp\{-(v_{12}-v_{13})^2/2[(\Delta v_{12})^2 + (\Delta v_{13})^2]\}$$

where $(\Delta v_{12})^2$ is the second moment of the absorption line between the levels 1 and 2, and $(\Delta_{13})^2$ is similarly defined. The importance of overlap in determining the magnitude of this transition probability is clearly demonstrated in the above expression.

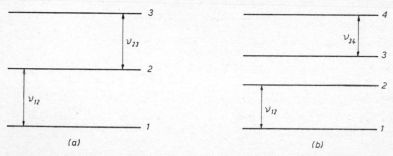

FIG. 3.2. Two other energy-level systems with strong probabilities of cross relaxation.

A cross-relaxation time T_{21} may be defined in such a case by the equation

$$(2T_{21})^{-1} = \sum_B w_{AB} p_B^{(13)}$$

The right-hand side of this equation represents the probability that a spin A will abosrb a quantum hv_{12} while an arbitrary spin B emits a quantum hv_{13}; $p_B^{(13)}$ represents the probability of the spin B being in the upper state of the 1–3 transition.

Bloembergen *et al.* considered other energy-level situations wherein pairs of levels occurred with similar but not identical separations. Examples are shown in Fig. 3.2, and it is noted that cross relaxation does not necessarily require the near coincidence of the energy levels. The important requirement is that the energy-level *separations* shall be similar. Thus in Fig. 3.2(*a*), appropriate to Ni^{2+} in an axial crystal field and a moderate magnetic field, the two similar frequencies are ν_{12} and ν_{23}. A further example is given by Fig. 3.2(b), the Cr^{3+} ion in an axial field where ν_{12} and ν_{34} can be nearly or exactly identical. In the latter case the levels 2 and 3 can be quite widely separated from one another and yet the cross relaxation may be most effective.

In the foregoing discussion, a 1 : 1 spin relation was postulated between the transitions occurring within the spin systems A and B, with ν_{12} similar to ν_{13}, as in the figures. This will be the most potent condition for cross relaxation generally, but the other cross-relaxation processes are also possible. In general, cross relaxation

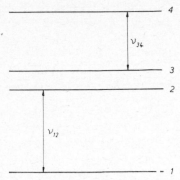

Fig. 3.3. Energy level diagram in which $a\nu_{12} = b\nu_{34}$ and a and b are small integers.

can occur with a finite transition probability with $(a+b)$ spins taking part, when a spins of type A flip down say, while b spins of type B flip up, so that

$$a(h\nu_{12}) = b(h\nu_{13}) \qquad (3.3)$$

The probability of this process occurring is clearly likely to be small unless a and b are integers close to unity. The process described by equation (3.3) has become known as *harmonic cross relaxation.*

If cross relaxation can occur, another route is thereby provided by which the populations of the spin levels can alter. Thus the rate equations must be modified to take this into account. We consider as an example the system of Fig. 3.3 where we will assume the condition

$$a\nu_{12} = b\nu_{34}$$

We further postulate the incidence of an external field of frequency ν_{12} and we define a cross-relaxation transition probability W_{CR}

such that it is the probability per unit time of a group of a spins making the $2 \to 1$ transition and b spins simultaneously making the $3 \to 4$ transition. A total of $(a+b)$ spins is involved in a single transition. Thus $W_{CR}(n_2)^a(n_3)^b$ is the number of transitions of this type occurring in unit time, each adding a spins to level 1 and removing b spins from level 3. A similar expression accounts for the transitions $1 \to 2$ and $4 \to 3$. Thus, writing $T_1^{(12)}$ for the spin-lattice relaxation time between levels 1 and 2, the rate equations become

$$\frac{d(n_1 - n_2)}{dt} = -\frac{1}{T_1(12)}\left[n_1 - n_2 - (N_1 - N_2)\right]$$

$$+ aW_{CR}(n_2{}^a n_3{}^b - n_1{}^a n_4{}^b) - 2W_S(n_1 - n_2)$$

$$\frac{d(n_3 - n_4)}{dt} = -\frac{1}{T_1(34)}\left[n_3 - n_4 - (N_3 - N_4)\right]$$

$$- bW_{CR}(n_2{}^a n_3{}^b - n_1{}^a n_4{}^b) \qquad (3.4)$$

Here W_S is the probability of a spin making the $1 \to 2$ or $2 \to 1$ transition in unit time in the presence of the external field of frequency v_{12}. The other symbols have their usual meaning.

We now make the important assumption that the separations of all four levels are small compared with kT so that

$$n_1 = n_c + \Delta_1$$
$$n_2 = n_c + \Delta_2$$
$$n_3 = n_c + \Delta_3$$
$$n_4 = n_c + \Delta_4$$

where the Δ's are small so that terms in Δ^2 may be ignored. This is a reasonable assumption in many experimental arrangements and will enable the effects of cross-relaxation to be shown in simple analytical form. Here n_c represents a constant population level.

Now consider the term

$$n_2{}^a n_3{}^b - n_1{}^a n_4{}^b = (n_c + \Delta_2)^a(n_c + \Delta_3)^b - (n_c + \Delta_1)^a(n_c + \Delta_4)^b$$

Expanding, this becomes

$$n_c{}^{a+b-1}[a(n_2 - n_1) + b(n_3 - n_4)]$$

We assume, for simplicity, that $T_1(12) = T_1(34) = T_1$ although this is not necessarily true. Writing $n_1 - n_2 = n_{12}$, $N_1 - N_2 = n_{012}$ the rate equations (3.4) become

$$\dot{n}_{12} = -\frac{(n_{12} - n_{012})}{T_1} + aW'(-an_{12} + bn_{34}) - 2W_S n_{12} \qquad (3.5a)$$

$$\dot{n}_{34} = -\frac{(n_{34}-n_{034})}{T_1} - bW'(-an_{12}+bn_{34}) \qquad (3.5b)$$

where $W' = W_{CR}n_c{}^{a+b-1}$.

If the relaxation between the pair of levels 1 and 2 is being studied for example by the pulse-saturation method (§7.1), the population difference is disturbed from the thermal equilibrium value and the recovery towards this value is then observed. During the recovery the saturating radio-frequency field is turned off and W_S is zero. The rate equations governing the recovery of both sets of populations are given by equation (3.5) with W_S set to zero, thus

$$_{12} = -\frac{n_{12}-n_{012}}{T_1} + aW'(-an_{12}+bn_{34}) \qquad (3.6a)$$

$$_{34} = -\frac{n_{34}-n_{034}}{T_1} - bW'(-an_{12}+bn_{34}) \qquad (3.6b)$$

Remembering that within our approximations the thermal equilibrium population differences are simply related by

$$an_{012} = bn_{034}$$

equations (3.6) may be solved to yield

$$\begin{aligned} n_{12}-n_{012} &= R_{12}\exp(-t/T_1)+R_{12}{}'\exp(-t/\tau) \\ n_{34}-n_{034} &= R_{34}\exp(-t/T_1)+R_{34}{}'\exp(-t/\tau) \end{aligned} \qquad (3.7)$$

where a cross-relaxation time T_{21} has been defined by

$$(a^2+b^2)W' = T_{21}{}^{-1}$$

and $$\tau^{-1} = T_1{}^{-1}+T_{21}{}^{-1}$$

By forming n_{12}, $\dot{n}_{34}$ from equation (3.7) and using equation (3.6) it may be shown that

$$\frac{R_{12}{}'}{R_{34}{}'} = -\frac{a}{b}, \quad \frac{R_{12}}{R_{34}} = \frac{b}{a}$$

Thus equation (3.7) becomes finally

$$n_{12}-n_{012} = R_{12}\exp(-t/T_1)+R_{12}{}'\exp(-t/\tau) \qquad (3.8a)$$

$$n_{34}-n_{034} = \frac{a}{b}R_{12}\exp(-t/T_1)-\frac{b}{a}R_{12}{}'\exp(-t/\tau) \qquad (3.8b)$$

Equations similar to these are quoted by Manenkov and Prokhorov[2] and by Feng[3] for slightly different cases. It is seen that even in the relatively simple case postulated, the recovery curve, which will be experimentally monitored to determine T_1, is no longer a single exponential in T_1 but is the sum of two exponentials, the second containing the cross-relaxation parameter T_{21}. It will be instructive to examine the form of the resulting recovery curve, and we do this for two extreme experimental situations.

(a) *Saturating pulse short compared with* T_{21}: By the time the pulse switches off, the 1–2 transition is saturated but the 3–4 transition has not been disturbed. Thus $n_{34} = n_{034}$ and in the recovery equation (3.8b), when $t = 0$

$$0 = \frac{a}{b} R_{12} - \frac{b}{a} R_{12}'$$

Thus the recovery of thermal equilibrium populations of levels 1 and 2 follows equation (3.8a) with

$$\frac{R_{12}'}{R_{12}} = \frac{a^2}{b^2}$$

The amplitudes of the two exponentials can thus be comparable and their separation, experimentally, depends upon there being a large difference between T_{21} and T_1.

(b) *Saturating pulse long compared with* T_{21}: It is important to detail the conditions assumed in the analysis. A high-power pulse is applied for a period sufficiently longer than T_{21} that no further changes in n_{12} and n_{34} occur. Thus $\dot{n}_{12} = \dot{n}_{34} = 0$ and n_{12} is reduced almost to zero. In applying equation (3.6b) the conditions are thus such that an_{12} can be neglected in comparison with bn_{34}. Hence

$$\dot{n}_{34} = 0 = -\frac{n_{34} - n_{034}}{T_1} - b^2 W' n_{34}$$

Remembering that $(a^2 + b^2) W' = T_{21}^{-1}$, we find

$$n_{34} \left(1 + \frac{b^2}{a^2 + b^2} \frac{T_1}{T_{21}}\right) = n_{034}$$

This is the value of n_{34} when $t = 0$ in equation (3.8b) and the recovery equations may be written, at that instant,

$$-n_{012} = R_{12} + R_{12}'$$

$$n_{034} \left\{ \left[1 + \left(\frac{b^2}{a^2 + b^2}\right) \left(\frac{T_1}{T_{21}}\right)\right]^{-1} - 1 \right\} = \frac{a}{b} R_{12} - \frac{b}{a} R_{12}'$$

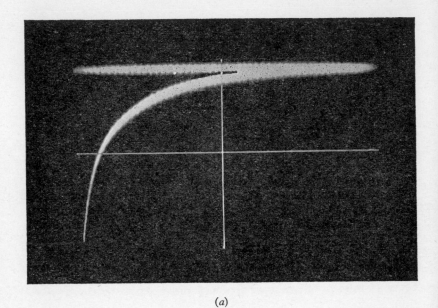

(a)

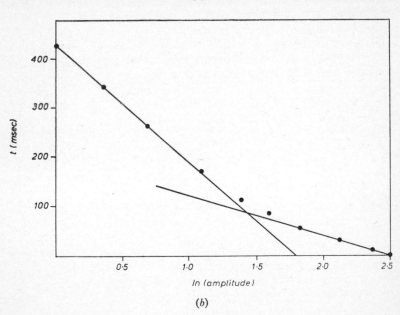

(b)

FIG. 3.4. The experimental two-exponential recovery curve in (a) is analysed into its two components in (b).

Further manipulation leads to

$$\frac{R_{12}'}{R_{12}} = \frac{a^2}{b^2} \frac{T_{21}}{T_1 + T_{21}}$$

Thus if a and b are comparable in magnitude and T_{21} is much shorter than T_1, as is often the case, after a short saturating pulse has been applied the recovery curve consists of a fast part followed by a slower one, of which an example is shown in Fig. 3.4. If the saturating pulse is very long, cross relaxation takes place during the pulse and, although the recovery still has two terms, the amplitude of that in τ is very much less than that in T_1 and the curve may appear like a single exponential, except in the initial part. In either case, so long as T_1 is appreciably greater than T_{21}, the final part of the recovery curve contains little of the τ (or T_{21}) amplitude; thus the derivation of T_1 from experimental data should be taken from this part of the curve whenever possible.

Experimental observation of cross relaxation

The work of Bloembergen et al.[1] and of Pershan[4] in 1960 included the first experimental observations of cross-relaxation effects which were identified as such. Pershan's investigation was in the field of nuclear magnetic resonance, and an early and most instructive investigation of cross-relaxation effects in electron paramagnetic resonance was reported by Mims and McGee[5] in 1960. They worked on ruby, in which the Cr : Al ratio ranged from 0·02 per cent to 0·3 per cent, and they examined the relaxation over a wide range of fields and angles with respect to the crystal axis. Their measuring frequency was 7·17 Gc/s. As indicated above, cross-relaxation effects appeared in the form of recovery curves having two or more different characteristic times. Mims and McGee also studied the effect of varying either the orientation or the magnetic field so that the precise 1 : 1 ratio of the two energy intervals was slightly disturbed.

Fig. 3.5 indicates schematically the four energy levels of ruby ($S = 3/2$) in an applied magnetic field. The six possible transitions are noted using the nomenclature which appears in Mims and McGee's paper. Fig. 3.6 shows the recovery curves when the relaxation of the transition B was studied in the neighbourhood of the setting for which $B = C$ (22° and 3·8 Gc/s). On either side of this precise position, the separation between B and C for this material increases at a rate of 300 Mc/s per degree. Decay traces close to 22° showed two distinct periods, the fast one corresponding to cross relaxation, the transfer of energy from the B to the C transition, and the slower one corresponding to their joint relaxa-

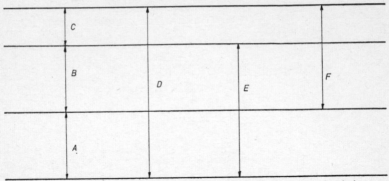

FIG. 3.5. The notation of Mims and McGee[5] for the energy levels in ruby. This is used also in the following four figures relating to their work.

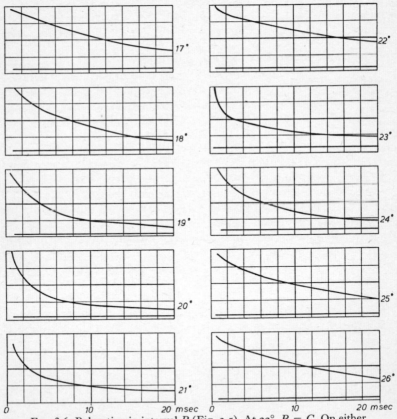

FIG. 3.6. Relaxation in interval B (Fig. 3.5). At $22°$ $B = C$. On either side of this setting cross relaxation results in a curve with two recovery parameters. Concentration of $Cr^{3+} = 0.1$ per cent; saturating pulse duration 0.15 msec. (After Mims and McGee[5].)

tion. The smaller the difference between the frequencies of B and C the faster was the observed rate of energy transfer by cross relaxation, until at 22° precisely transfer took place during the interval of the saturating pulse and was already complete by the time the observation of the recovery could begin. Thus in Fig. 3.6 the curve at 22° does not show the fast cross relaxation which is apparent in the recoveries just on either side of this angle. Even when the intervals B and C were separated by several hundred megacycles per second, cross relaxation was rapid compared with the true spin-lattice relaxation. When the difference between the intervals B and C was 600 Mc/s, the cross-relaxation time was still 3 msec for a chromium concentration of 0·1 per cent. This time was unaffected by changing the temperature from 4·2°K to 20°K, as would be expected for a spin-spin phenomenon; it was, however, very sensitive to concentration. At 0·15 per cent Cr^{3+} and with the same B and C separation, the cross-relaxation time shortened to 1 msec whereas at 0·02 per cent no cross relaxation could be detected at such large frequency separations. It is relevant to record that the linewidth of the ruby absorption is of the order of 50 Mc/s, so that a separation of 600 Mc/s really involves overlap of the two absorption curves in the wings only.

TABLE 3.1. Three spin cross-relaxation times (in milliseconds).

Interval relation	Observation interval	Cr/Al concentration (per cent)				
		0·3	0·15	0·1	0·05	0·02
$B = 2A$	A	0·08	0·4	1·2	4	50
$C = 2A$	A	0·06	0·35	1	4	50
$A = 2B$	B	0·08	0·4	1·3	4·5	70
$E = 2C$	C	—	0·2	0·7	2·8	50
$C = 2B$	C	—	0·14	0·4	1·5	40
$A = 2C$	C	0·05	0·18	0·6	2·4	55
$B = 2C$	B	0·07	0·5	1	3·5	—
$D = 2B$	C	—	—	0·2	0·6	13

The B to C transition referred to above was a 1 : 1 process. Mims and McGee worked also at an angle of 74° and 3·55 Gc/s when the interval C is twice that of A, the difference between C and $2A$ being 570 Mc/s per degree on either side of the exact setting. The cross-relaxation time was found in this case with three spins taking part and it was much longer than in the previous case with the two intervals identical. Moreover, the effect could not be seen unless the condition $C = 2A$ was very nearly obeyed. The traces in Fig. 3.7 were obtained by Mims and McGee with a sample containing 0·1 per cent Cr^{3+} at a temperature of 4·2°K, and

the cross-relaxation time was of the order of 1 msec. Measurements
were also made at several of the intervals shown in Fig. 3.5 and the
results for the various cross-relaxation times are shown in Table
3.1 where the intervals have the same nomenclature as that given
in the Fig. 3.5. Mims and McGee noted that the cross-relaxation

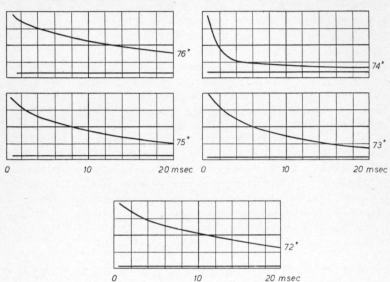

Fig. 3.7. Relaxation in the interval A (Fig. 3.5). At $74°$, $C = 2A$.
Cross relaxation occurs with a time constant of 1 msec. Concentration
0·1 per cent; saturating pulse duration 0·1 msec. (After Mims and
McGee[5].)

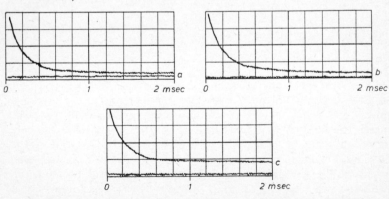

Fig. 3.8. General cross relaxation in 0·3 per cent ruby in the interval
A (Fig. 3.5) at $80°$ to the crystal axis. Cross-relaxation time constant
0·14 msec; T_1(d) = 30 msec; saturating pulse durations (a) 0·05,
(b) 0·3, (c) 1·0 msec. Temperature $4·2°$K. (After Mims and McGee[5].)

time for each of these intervals obeyed roughly the same power law of variation with concentration, viz (concentration)$^{2\cdot4}$.

The above data were obtained when close to an orientation such that one interval bore a simple relation to another. At a concentration of 0·3 per cent Cr^{3+}, however, two period recovery curves were observed by Mims and McGee at all angles and field settings. Fig. 3.8 shows the observed relaxation at interval A and an angle of 80°, a frequency of 4·05 Gc/s and at 4·2°K. The initial part of the recovery has a time constant of 0·14 msec while the later slow recovery has a 30 msec period. The relative proportions of the curve occupied by these two recoveries depended upon the duration of the saturating pulse in the manner shown. At this orientation, there is no simple relation to any of the other three frequency intervals. At different settings, fast decay times were found ranging from 0·04 to 0·3 msec. This general fast decay for a given setting was unchanged in its value when the temperature was altered from 4·2 to 2·2°K, behaving thus in the manner described above and reinforcing the suggestion that it is due to the transfer of energy between the pumped transition and other intervals in the level scheme. This type of behaviour is usually found only in samples containing quite a high concentration of magnetic ions, and it is referred to as *general cross relaxation*.

Mims and McGee drew attention to the problem of interpreting recovery curves when the cross-relaxation and spin-lattice relaxation times are of the same order of magnitude. This can readily occur in relatively dilute specimens when the operating point is slightly moved from that corresponding precisely to a 1 : 1 cross-relaxation point. The example given by these authors is interesting. A transition C at 90° and 1·4 Gc/s was examined. With a crystal containing 0·3 per cent Cr^{3+} the fast component at 4·2°K had a period of 0·04 msec and the slow component 30 msec. At 0·15 per cent the fast period was 1·1 msec and the slow period 100 msec, so that the two could still be seen separately. With a specimen of 0·1 per cent Cr^{3+} there was a recovery curve with an apparently single exponential of time constant 5 msec which could readily be mistaken for a true spin-lattice relaxation time. Further experiments, however, showed that this time was almost independent of temperature, varying only from 6 to 4 msec in the range 2·2 to 20°K. The decay trace was not a good exponential and at the end of it a long tail could just be distinguished. A 0·05 per cent specimen gave recovery times which were longer than 5 msec but were still of indeterminate form being nearly exponential, while at 0·02 per cent the decay trace was a fair exponential with a time constant of approximately 500 msec at 4·2°K. The importance of this obser-

vation by Mims and McGee cannot be over stressed since it
indicates the ease with which errors can be made in relaxation
measurements if care is not taken to distinguish between cross-
relaxation effects and true spin-lattice relaxation. To emphasize
this, it is only necessary to observe that Geusic[6] has noted the
effect of harmonic cross-relaxation processes up to 5 : 1, in ruby,
with indications of effects as high as 11 : 1. It is believed that cross

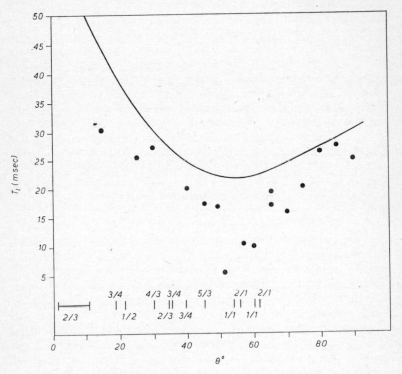

FIG. 3.9. Angular variation of $T_1(d)$ in ruby at 4·2°K. Experimental
points are full circles and the full line represents the calculated results
of Donoho, reduced by a factor of 3. The figures a/b refer to possible
cross relaxations of the form $a \times 35 = b \times v_{CR}$ (Gc/s) with $a+b$ spins
involved in the process. (After Lees *et al.*[8])

relaxation may at least partly account for the disagreement between
experiment and theory in the case of relaxation in ruby at 35 Gc/s.
The measurements of Standley and Vaughan[7] on certain ruby
crystals were in excellent agreement with the theoretical predictions of Donoho. These measurements were made at 9·3 Gc/s but
when the same crystals were used in measurements at 35 Gc/s, the
agreement was most unsatisfactory,[8] as Fig. 3.9 shows. This figure

also indicates the many harmonic cross-relaxation points at this higher frequency (and therefore higher fields) and these are operative at many angles. It is easy to see that such cross relaxation could account for the general reduction in the level of the observed relaxation time.

So far it has been tacitly assumed that cross relaxation takes place between similar ions, for the need to conserve or nearly conserve the Zeeman energy would appear to indicate this. The assumption is not necessary. If a crystal has two types of ions, say C and D, their absorption frequencies at a given field strength will generally not coincide. However, suppose that the C ions have very long spin-lattice relaxation times while the D ions relax quite rapidly. Then the width of the absorption line of the D ions can be so broadened that it overlaps the narrow line of the C ions over a wide range of angles and fields. The D ions are therefore able to absorb at the resonant frequency of the C ions and they can emit that same frequency also. But this is just the frequency needed to induce transitions in the C ions, so cross relaxation of a particular C ion into the wings of a neighbouring D ion becomes possible. For example, it is now thought that some of the unexpected results attributed to the Cr^{3+} ion in certain samples of ruby (see Chapter 9) may be due to the presence of some of the chromium in the divalent state, and this would then constitute the D ion above. Again, it is possible that the doping of such a crystal may result in the formation of clusters of ions. The relaxation times of a cluster can be shorter than that of an isolated ion, for a cluster may be able to relax by processes which the single ion cannot utilize. (Gill,[9] for example, has shown that exchange forces within the cluster will be modulated by thermal vibrations in the lattice and can give a powerful relaxation mechanism.) Thus a cluster of ions, a pair or more ions, could be the D ion above and cause the observed relaxation of the ion C to be considerably modified.

Use of cross relaxation in harmonic pumping

In the solid-state maser oscillator or amplifier there are generally three or more energy levels involved. A high-frequency pump power is applied so as to saturate a given pair of levels and the requirement is that such saturation will invert the populations of another pair of levels. For example, in Fig. 3.10, a signal applied between levels 1 and 4 so as to saturate them will produce a population of level 4 greater than that in level 3. Thus, if the 3–4 transition is stimulated, the possibility occurs of maintaining an oscillation at this frequency by continuous saturation of the

E

1–4 transition. This is a very brief and inadequate description of the principle of operation of the solid-state maser, but it suffices here for our purpose. If further information is required, there are many publications on this topic, for example Siegman.[10]

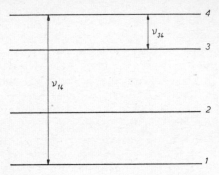

Fig. 3.10. Possible three-level maser energy levels. If power at frequency ν_{14} saturates the transition, the population of level 4 may exceed that of level 3.

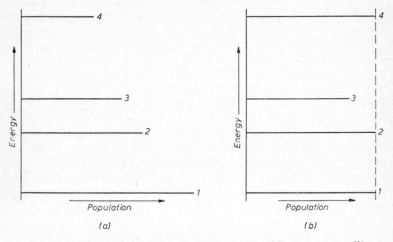

Fig. 3.11. (a) Postulated energy-level system with $2\nu_{12} = \nu_{24}$. (b) Inversion of the 3–4 transition, by direct saturation of levels 1 and 2, with cross-relaxation saturation of levels 2 and 4.

It is of interest here to note that one may use harmonic cross relaxation in order to bring about the inversion of the population required for maser action. There are a number of situations where this can occur and the example given below is not intended to be more than an illustration. Consider the energy-level diagram given in Fig. 3.11(a). Here, the lengths of the horizontal lines have been drawn roughly in proportion to the populations which one would expect when such a system were in thermal equilibrium, with the assumption of all energy separations being small compared with kT. We will now suppose, as an example, that v_{12} is harmonically related to v_{24} and in the figure we have shown this as 1 : 2. What happens when we apply a saturating pulse of frequency v_{12}? Assume that the spin-lattice relaxation time for all the levels is quite long and that there is a rapid cross relaxation between the 1-2 and 2-4 pairs of levels. If we now have a saturating pulse of adequate power, levels 1 and 2 will be almost saturated directly, while levels 2 and 4 will also be almost saturated, by cross relaxa-

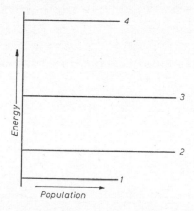

FIG. 3.12. Maser cross-relaxation scheme used by Arams.[11] A pump signal saturates levels 2 and 3 directly and levels 1 and 4 by harmonic cross relaxation. A population in level 3 greater than in level 1 was produced in this way.

tion. Under the simplifying assumptions which we have made, the populations of levels 1, 2 and 4 will be approximately equal as shown in Fig. 3.11(b). So far we have not disturbed the population of level 3 and if we can arrive at the state of equal populations in levels 1, 2 and 4 without at the same time greatly increasing the population of level 3, we can achieve the condition required for successful stimulation of the oscillation between levels 4 and 3, that is n_4 greater than n_3. Working with ruby as the operating material in a field close to 1700 gauss near $\theta = 90°$, Arams[11] used the situation shown diagrammatically in Fig. 3.12. He was able to saturate the transition 1–4 by cross relaxation while applying the pump frequency between levels 2 and 3. He was then able to get maser amplification using the 1–3 transition.

An experimental study of the 2 : 1 harmonic cross relaxation in ruby was carried out by Squire.[12] His aim was to discover the rate, linewidth and line-shape of the harmonic cross relaxation, with the ultimate intention of producing a millimetre-wave maser. This was not achieved but his results are nevertheless most interesting. His technique was to saturate a transition at 9·6 Gc/s and to observe the effect of this on the linewidth of a second transition at 19·2 Gc/s. Four operating points were available in ruby where the required frequencies were to be found and these are shown in Fig. 3.13. Squire's work gives a direct measure of the cross-

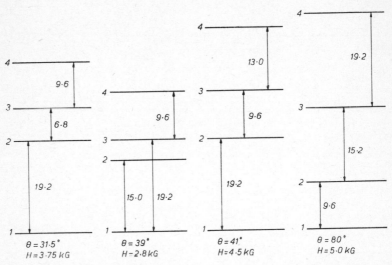

FIG. 3.13. The four ruby operating points used by Squire,[12] showing the energy levels involved in the 2 : 1 harmonic cross relaxation. The level separations are given in Gc/s.

relaxation probability for a system of this kind and also of the linewidth. He obtained results which were slightly different for the different operating points but it suffices to say that the linewidth lay between 100 and 200 Mc/s for cross relaxation at 4·2°K and this did not change when the operating temperature was raised to 20·3°K. He obtained a value for the cross-relaxation transition probability of 110 sec⁻¹ for a sample containing 0·028 per cent chromium, while Bogle and Gardner[13] obtained a value of 700 sec⁻¹ for a sample containing 0·013 per cent chromium. The difference between these two values is probably not significant since it could be accounted for by differences in the purities of the ruby samples.

It is interesting that the first maser operation reported by Feher and Scovil[14] using a mixed crystal of lanthanum ethyl sulphate containing gadolinium, which has a spin S of $\frac{7}{2}$ and cerium, for which $S = \frac{1}{2}$. Feher and Scovil operated this material at a particular applied magnetic field value and orientation so that one of the

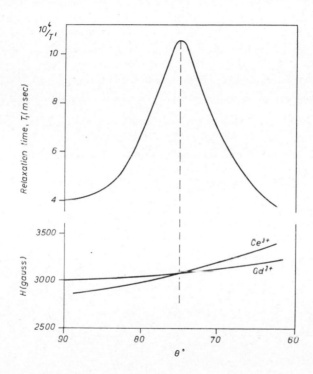

Fig. 3.14. Cross relaxation between cerium and gadolinium ions in lanthanum ethyl sulphate. The upper curve shows the relaxation rate as a function of orientation for one gadolinium transition while the lower curve shows the resonance fields. (From the data of Feher and Scovil.[14])

gadolinium transition frequencies was exactly equal to that of the cerium. Thus there was a 1 : 1 cross-relaxation process between ions of different types. The spin-lattice relaxation rate of the Ce^{3+} ions was appreciably faster than that of the Gd^{3+} ions and Fig. 3.14 shows the effect of the cerium on the measured gadolinium relaxation time.

A similar effect has been shown by Standley and Vaughan.[15] They were searching for an effect on the relaxation time of the Cr^{3+} ion in ruby due to the known presence of Cr^{4+} ions in a particular sample. As Fig. 3.15 shows, the effect of cross relaxation between different pairs of energy levels in the Cr^{3+} was much

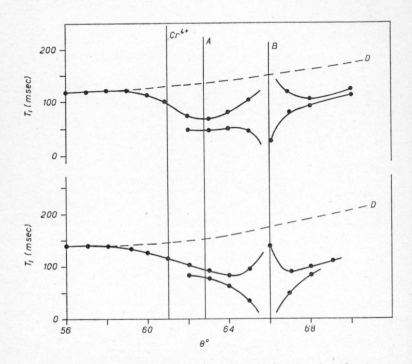

FIG. 3.15. The angular variation of measured relaxation times for the Cr^{3+} (2–3) transition in ruby at $4.2°K$. The upper curves relate to orange ruby, containing also Cr^{4+}, the lower to pink ruby. The dashed line represents Donoho's calculated results. At 'Cr^{4+}', the Cr^{3+} and Cr^{4+} resonances coincide in field. At the harmonic point A, $2\nu_{23} = 3\nu_{12}$; at B, $V_{23} = \nu_{34}$ referring to the four Cr^{3+} spin levels. (After Standley and Vaughan.[15])

more effective than the cross relaxation between Cr^{4+} and Cr^{3+}. The angular extent of the harmonic cross relaxation is clearly shown in this figure.

Experimental observation of cross relaxation, harmonic and otherwise, has been well reported and for further details the reader is referred, for example, to Siegman[10] where an extensive bibliography will be found.

References

1. Bloembergen, N., Shapiro, S., Pershan, P. S., and Artman, J. O., *Phys. Rev.*, 1959, **114,** 445.
2. Manenkov, A. A., and Prokhorov, A. M., *Sov. Phys. JETP.*, 1962, **15,** 54.
3. Feng, S., *Cruft Lab. Tech. Rept.*, **371** (Harvard University, 1962).
4. Pershan, P. S., *Phys. Rev.*, 1960, **117,** 109.
5. Mims, W. B., and McGee, J. D., *Phys. Rev.*, 1960, **119,** 1233.
6. Geusic, J. E., *Phys. Rev.*, 1960, **118,** 129.
7. Standley, K. J., and Vaughan, R. A., *Phys. Rev.*, 1965, **139,** A1275.
8. Lees, R. A., Moore, W. S., and Standley, K. J., *Proc. Phys. Soc. (Lond.)*, 1967, **91,** 105.
9. Gill, J. C., *Proc. Phys. Soc.* (Lond.), 1962, **79,** 58.
10. Siegman, A. E., *Microwave Solid-State Masers* (McGraw-Hill Inc., 1964).
11. Arams, F. R., *Proc. I.R.E.*, 1960, **48,** 108.
12. Squire, P. T., *Proc. Phys. Soc. (Lond.)*, 1965, **86,** 573.
13. Bogle, G. S., and Gardner, F. F., *Aust. J. Phys.*, 1961, **14,** 381.
14. Feher, G., and Scovil, H. E. D., *Phys. Rev.*, 1957, **105,** 760.
15. Standley, K. J., and Vaughan, R. A., *Proc. Phys. Soc. (Lond.)*, 1965, **86,** 861.

Chapter 4

The Phonon Bottleneck

In the preceding chapters our discussion of the relaxation phenomena was based upon the assumption that the lattice and bath
temperatures are identical, and remain so during an experiment.
Putting this another way, the spins transfer their energy to phonons
which are assumed to be in such excellent communication with the
surrounding bath that the phonon excitation always has its value
P (see p. 11) appropriate to the bath temperature T_b. Clearly the
calculated spin-lattice relaxation times are strictly *spin-phonon*
relaxation times while experiments measure *spin-bath* relaxation
times. Under our assumptions, experiment and theory effectively
deal with identical quantities, but will this always be so? In the
Raman-type relaxation a wide band of the phonon spectrum is
accessible and there should be no difficulty generally in meeting
the requirements of the above assumptions. A different situation
can obtain at lower temperatures as we shall now show.

Considering the direct process, a single spin flip is accompanied
by the creation of a single phonon of which the frequency is equal
to the resonant frequency of the E.S.R. transition (or to the frequency v of the incident microwave power which is assumed to
be the same). Relatively low phonon energies are thus involved,
occupying a small frequency band Δ_v around v, where Δ_v is the
phonon-spin linewidth which is not necessarily the same as the
photon-spin linewidth. Since there are relatively few phonons
available to take part in the direct relaxation process, there must
exist the possibility of the phonons being excited to a higher energy
state, corresponding to a temperature greater than T_b. In other
words, in these circumstances the spin specific heat may be very
much larger than the phonon specific heat with consequent rise in
temperature of the phonons, at least within this limited frequency
band. Any effect due to this rise in temperature will only be
observed—indeed the rise in phonon temperature will only in a
sense be real—if the phonon lifetime in question is relatively long.
A very short lifetime will imply a rapid transfer of energy from
this narrow band of phonons through the rest of the phonon

spectrum and to the surrounding bath. The shortage of available phonons creates what is called a phonon bottleneck, a phenomenon first discussed by Van Vleck[1] in 1941. We shall now examine some of its consequences.

We follow Faughnan and Strandberg[2] in deriving the spin-phonon-bath equations. We note that the number of phonons in the frequency band Δ_v is $\rho(\delta)P(\delta)\Delta(\delta)$, where ρ is the density of states in the frequency band Δ_v and is given classically by equation (1.12)

$$\rho(\delta)d(\delta) = \frac{3V\delta^2}{2\pi^2\hbar^3 v^3}\,d(\delta)$$

where V is the volume of the sample, v is the velocity of sound, and $\delta = hv$. The phonon excitation number P is given by equation (1.13) (p. 11) as

$$P(\delta) = [\exp(\delta/kT)-1]^{-1}$$

in thermal equilibrium. We make the assumption that if this number is disturbed from its equilibrium value the return to equilibrium is exponential, characterized by a phonon relaxation time T_{ph}. We designate as p the non-equilibrium value of P. Then recalling that a spin flip alters n by two units but the number of phonons by only one we have

$$\frac{d}{dt}(\rho p h \Delta_v) = \frac{\dot{n}}{2} - \frac{\rho h \Delta_v (p-P)}{T_{ph}}$$

or

$$\frac{dp}{dt} = \frac{\dot{n}}{2\rho h \Delta_v} - \frac{(p-P)}{T_{ph}} \tag{4.1}$$

Similarly, using the arguments of Chapter 1, pp. 10–12, the rate equation for the spin system becomes

$$n = -T_1(\mathrm{d})^{-1}\tanh(\delta/2kT)\,[n(2p+1)-N] \tag{4.2}$$

where $N = N_a + N_b$.

The equations (4.1) and (4.2) are the equations for the coupled spin-phonon system and reduce to the standard form considered earlier if p has the equilibrium value P at all times.

Since $n_0/N = \tanh(\delta/2kT)$, equation (4.2) becomes

$$\dot{n} = -\frac{1}{T_1(\mathrm{d})}\frac{n_0}{N}\,[n(2p+1)-N]$$

where n_0 indicates the thermal equilibrium value. Equations (4.1) and (4.2) may be rewritten in the form

$$\frac{dp}{dt} = -\frac{1}{T_{ph}} \, \sigma \left(\frac{P+\frac{1}{2}}{n_0}\right) (n-n_0) - \frac{1}{T_{ph}} \left(\sigma \, \frac{n}{n_0} + 1\right) (p-P)$$

$$\tag{4.3}$$

$$\dot{n} = \frac{n-n_0}{T_1(\mathrm{d})} - \frac{n}{T_1(\mathrm{d})} \left(\frac{p-P}{P+\frac{1}{2}}\right) \tag{4.4}$$

The parameter σ is called the bottleneck factor and is given by

$$\sigma = \frac{T_{ph}n_0}{2T_1(\mathrm{d})(P+\frac{1}{2})(\Delta\delta)\rho(\delta)}$$

It may also be written as

$$\sigma = \frac{E_z/T_1(\mathrm{d})}{E_{ph}/T_{ph}}$$

where E_z is the Zeeman energy of the spin system $[E_z = \frac{1}{2}N\delta \times \tanh(\delta/2kT)]$ and E_{ph} is the energy of the phonon system $[E_{ph} = \frac{1}{2}\rho(\delta)h\Delta_v \coth(\delta/2kT)]$. Thus σ is the ratio of the energy exchange rate between spins and phonons with respect to the energy exchange rate between phonons and bath. If $\sigma \ll 1$, no bottleneck exists; if $\sigma \gg 1$ we have severe bottleneck conditions.

Scott and Jeffries[3] note that measurements are frequently made when recovery of the thermal equilibrium of the spin system is almost complete. Then $n \approx n_0$ and equations (4.3) and (4.4) become linearized thus

$$\left.\begin{aligned}\frac{dp}{dt} &= -\frac{1}{T_{ph}} \, \sigma \left(\frac{P+\frac{1}{2}}{n_0}\right) (n-n_0) - \frac{1}{T_{ph}} (\sigma+1) (p-P) \\[2mm] \dot{n} &= \frac{n-n_0}{T_1(\mathrm{d})} - \frac{n_0}{T_1(\mathrm{d})} \left(\frac{p-P}{P+\frac{1}{2}}\right)\end{aligned}\right\} \tag{4.5}$$

The solutions of these two equations involve two time constants T_a and $T_a{}'$ where

$$T_a \simeq T_1(\mathrm{d}) + \tau_1$$

$$\frac{1}{\tau_1} = \frac{1}{T_{ph}} \frac{E_{ph}}{E_z} = \frac{1}{T_{ph}} \frac{3\delta^2\Delta(\delta) \coth^2(\delta/2kT)}{2\pi^2cv^3\hbar^3}$$

$$\approx \frac{1}{T_{ph}} \frac{6(\Delta\delta)k^2T^2}{\pi^2cv^3\hbar^3} \quad \text{for } \delta \ll 2kT$$

$$= DT^2$$

$$\frac{1}{T_a{}'} = \frac{1}{T_1(\mathrm{d})} \frac{E_z}{E_{ph}} \approx \frac{A\pi^2cv^3\hbar^3}{6\Delta(\delta)k^2T} = \frac{F}{T}$$

and c is the number of spins per cc.

It should be mentioned that these solutions assume $E_z \gg E_{ph}$ and that either $\sigma \gg 1$ or $T_1(\mathrm{d})$ and T_{ph} are very different.

Although n and p each decay in a manner involving both time constants T_a and T_a' it may be inferred that, in general, the main change in n occurs in a time T_a while p changes in a much shorter time T_a'.

Scott and Jeffries show that when T_a' is so short that the *observed* recovery curve is characterized by the longer T_a, then since $T_1(\mathrm{d})^{-1} = AT$ and $\tau^{-1} = DT^2$

$$\frac{1}{T_a} = \frac{ADT^3}{DT^2 + AD} \rightarrow AT \quad \text{if } DT^2 \gg AT \text{ (no bottleneck)}$$

$$\rightarrow DT^2 \quad \text{if } DT^2 \ll AT \text{ (severe bottleneck)}$$

The above analysis reveals the first important difference to be expected when there is a distinct phonon bottleneck at low temperatures. The measured relaxation times from the tail of the recovery curves will follow a temperature dependence of T^2 rather than T. If we analyse the above equations under the assumption that T_{ph} is very much less than $T_1(\mathrm{d})$ we get a further interesting observation on the form of the recovery. Under these circumstances, we can derive an approximate expression for $(p-P)/(P+\frac{1}{2})$ from equation (4.3) and if this is then fed into equation (4.4), this can be solved to give a recovery curve of the form

$$\frac{t}{T_1(\mathrm{d})} = (1+\sigma) \ln \left(\frac{n_0}{n_0 - n} \right) - \frac{\sigma n}{n_0} \tag{4.6}$$

Clearly this equation reduces to the standard form if σ is zero. But equally obviously, if σ is finite and of such value that the last term in equation (4.6) cannot be ignored, the recovery is far from exponential. The assumption we have made in deriving this equation [that T_{ph} is very much less than $T_1(\mathrm{d})$] is not an unreasonable one for, in the example cited below, $T_1(\mathrm{d})$ was of the order of milliseconds whereas T_{ph} was of the order of 10^{-8} seconds.

The rate equation of which (4.6) is a solution can be modified if n is very nearly equal to n_0, that is to say, recovery is almost complete. The equation becomes

$$\dot{n} = -\frac{n - n_0}{(1+\sigma)T_1(\mathrm{d})} \tag{4.7}$$

which indicates that the experimental value of the recovery parameter is given by

$$T_{1e} = (1+\sigma)T_1(\mathrm{d}) \tag{4.8}$$

From the definition of σ above, it follows that σ is proportional to T^{-1} when $\delta \ll kT$ and therefore, from equation (4.8), when $\sigma \gg 1$ we have

$$T_{1e} \propto T^{-2}$$

This is the temperature dependence already shown for τ_1.

A theoretical analysis of the phonon bottleneck is given by Stoneham.[4] He uses a thermodynamic approach and includes in his analysis both the direct and the Orbach processes. He draws attention to the importance of the phonon relaxation time, the relative magnitude of which controls the situation completely, determining whether there shall be an apparent phonon bottleneck or not. Within the limit of the approximations made, Stoneham's results are similar to those above, but are more detailed in that the phonon relaxation process itself is dealt with at some length. We give an outline of his findings in the next section.

Phonon relaxation times

Energy can be transferred from the phonons responsible for relaxation to the constant temperature bath by several processes, in which the rate determining process may involve the transfer of energy in space or in frequency. Stoneham describes these respectively as (*a*) a spatial bottleneck for which the bath is the liquid helium surrounding the crystal and (*b*) a spectral bottleneck for which the lattice vibrations with energies of the order of kT form the bath.

In the case of a spatial bottleneck we consider a crystal whose dimensions are of order L and wherein phonons can reach the bath surrounding the crystal if $v\tau_c$, a phonon mean free path, exceeds L. τ_c is the phonon relaxation time found in thermal conductivity theory (Klemens[5]) and is not simply related to T_{ph} which describes the relaxation of phonons with a given energy. For a phonon mode resonant with the spins, the relaxation corresponding to τ_c is usually dominated by the elastic scattering of phonons by the spins from one resonant mode to another. Stoneham notes that in this case, for the direct process phonons of energy δ, we have

$$\tau_c(\delta) = \frac{\rho T_1(d)}{N} \left(\frac{2kT}{\delta}\right)^2$$

On the assumption that the interface between the crystal and the bath does not reflect phonons, Stoneham calculates T_{ph} using black-body radiation theory and finds

$$T_{ph} = \frac{4 \text{ (Volume of crystal)}}{v \text{ (Surface area of crystal)}}$$

which is $4L/3v$ for a sphere of radius L and $4L/v$ for a plate of thickness $2L$. v is the velocity of sound in the material which is assumed to be isotropic. In the former case, we have for the direct process in which spatial bottleneck of phonons occurs

$$T_{1e} = T_1(d) \left(1 + \frac{4}{3}\frac{L}{v\tau_c}\right) \qquad (4.9)$$

A similar type of equation holds for the Orbach case. Thus when the crystal radius is large compared with the phonon mean free-path, the bottleneck is particularly severe. When the bottleneck is slight these equations are not truly valid, but it is clearly apparent that there will be a dependence of T_{1e} upon the dimensions of the crystal. This point will be mentioned again below in connection with the experimental evidence for the phonon bottleneck.

When L is large it is not a good assumption that the temperature of the lattice is uniform over the crystal. This case has been treated by Eisenstein[6] who finds, for the temperature region where the direct process of relaxation is predominant, that the phonon relaxation time is proportional to L^2. It should be noted that the bath need not be the liquid helium surrounding the crystal, but may be impurities, defects or cracks which transfer the energy to some heat sink so rapidly that the rate determining process is the transfer of energy in space. In these cases, L may be related to the mean separation of the defects in question and is not then one of the external dimensions of the crystal. It is important to bear this in mind when considering the experimental evidence for the existence of the bottleneck based upon variation of T_{1e} with crystal size (see p. 71). In these processes energy is transferred from the phonon modes resonant with the spins to other thermal modes. Stoneham considers particularly the case where impurities play a major role. The main requirements of the impurity are a strong interaction with the phonon modes resonant with the spin system and a strong interaction with the thermal modes at the temperatures of the experiment. He shows that in the single-phonon process, for which the energy-level separation δ is close to an energy splitting of the impurity,

$$T_{ph}^{-1} \propto f_1 T^{-1}$$

where f_1 is the concentration of the impurities. If a two-phonon process is considered in which δ differs from an energy splitting of the impurity by about kT, then

$$T_{ph}^{-1} \propto f_1$$

and independent of temperature. Stoneham also considers the

effect of anharmonic processes which result in the energy transfer from the phonon modes resonant with the spins to the thermal modes and he obtains a relaxation time of the form

$$T_{ph}^{-1} \propto \delta T^4$$

Experimental observation of the phonon bottleneck

A good indication of the presence of the phonon bottleneck and of the validity of equation (4.6) above was provided by the experiments of Standley and Wright[7] on relaxation of the copper ion in copper dipyrromethene. These measurements were made

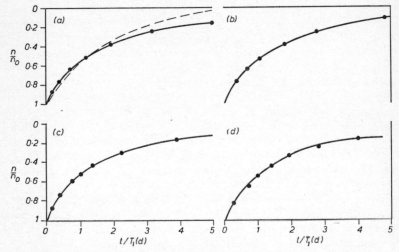

FIG. 4.1. Recovery curves for copper dipyrromethene (Standley and Wright[7]). Circles are experimental points while the full curve is calculated from equation (4.6). In (a) the broken curve is an exponential through $n/n_0 = 0.5$ and 1.0. Other data, given in the order T, σ, $T_1(d)$, are (a) $1.45°$K, 2.65, 1.64 msec; (b) $1.85°$K, 2.05, 1.33 msec; (c) $2.14°$K, 1.85, 1.06 msec; (d) $2.45°$K, 1.57, 0.94 msec.

at 9.2 Gc/s using a pulse saturation technique (§7.1) and they extended from 1.4 to $4.2°$K, although bottleneck conditions only occurred below $2.5°$K. The non-exponential behaviour of the recovery shown by equation (4.6) was clearly demonstrated (see Fig. 4.1). The spin system in this material was relatively easily saturated and it was possible to follow almost the entire recovery from $n = 0$ to $n = n_0$. The procedure was as follows. The experimental curve for the measurements at $1.45°$K was fitted at two points to the theoretical curve of equation (4.6) yielding values of σ and $T_1(d)$ at this temperature. Fig. 4.1 shows the resulting fit which was considered good. Values of σ were then calculated at other experimentally-used temperatures assuming that σ was proportional to T^{-1}. Using these calculated values the experimental and theoretical curves were fitted at one point only, namely $n/n_0 = 0.5$ to yield values of $T_1(d)$ at each temperature. The satisfactory delineation of the experimental results is shown in the

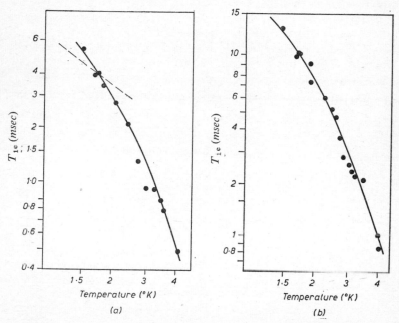

FIG. 4.2. Graphs of measured relaxation times T_{1e} against temperature for two copper dipyrromethene specimens. In (a) the broken line represents $(T_{1e})^{-1} = $ constant $\times\ T$. The equations of the full curves are

$$(a)\ (T_{1e})^{-1} = 82T^2 + 3.3 \times 10^{-3}T^9 \text{ and}$$
$$(b)\ (T_{1e})^{-1} = 36T^2 + 1.65 \times 10^{-3}T^9$$

with T_{1e} in seconds and T in °K.[7]

figure, wherein there is also indicated the degree of departure of the recovery from the simple exponential which one would expect were there no phonon bottleneck. The values of $T_1(d)$ calculated followed the expected inverse temperature law. Two different specimens gave similar but not identical recovery parameters, and these are indicated in Fig. 4.2. The T^9 dependence at high temperatures is expected for Raman-like behaviour in these ions and the T^2 dependence of the phonon bottleneck is very satisfactorily shown at lower temperatures.

Although these measurements subjected equation (4.6) to a good test, they were not extensive since bottleneck conditions were only found at temperatures below $2\cdot5°$K and values of σ were not high, $2\cdot65$ being the highest value determined, at $1\cdot45°$K.

More conclusive evidence for the bottleneck has been found by Taylor[8] in samples of the material di(tetrabutylammonium)bis (maleonitriledithiolato) Cu II, $[(C_4H_9)_4N]_2Cu[S_2C_2(CN)_2]_2$. It was possible to study relaxation at a particular orientation where the hyperfine splitting was so small that only a single slightly broadened resonance line was observed. A phonon bottleneck was found to exist from $1\cdot5$ to $4\cdot2°$K, the highest experimentally avail-

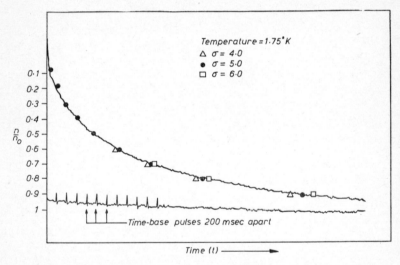

FIG. 4.3. Taylor's[8] data comparing an experimental recovery curve for CuMNT (see text) with points calculated from equation (4.6) for three values of σ.

able temperature in the liquid helium range. Taylor used a digital memory oscilloscope to improve the signal-to-noise ratio of his recovery curves and he was able to check the sensitivity of the fit of equation (4.6) to changes in the values of σ, which were as high as $5\cdot0$ at $1\cdot75°$K. As Fig. 4.3 shows, only in the tail of the recovery curve was the fit really sensitive to changes in σ, and the values of σ determined by curve fitting are not therefore very accurate in themselves. Taylor's data, however, may be said to provide a searching test of equation (4.6) which satisfactorily represents the experimental results.

In testing the assumptions of the foregoing theory, Taylor used both equations (4.6) and (4.8) to calculate the ratio $T_{1e}/T_1(d)$. He found agreement between the two only when σ was greater than about 3. He was also able to check the temperature variation of σ and found a $1/T$ dependence over only part of the experimental range (Fig. 4.4). He also investigated the variation in T_{1e} from sample to sample, in order to establish a possible size effect

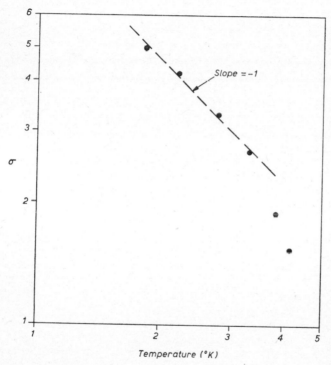

Fig. 4.4. Experimental dependence of σ on temperature for CuMNT (after Taylor[8]).

(equation 4.8). A variation was discovered without any obvious dependence on crystal size. The largest crystal had one of the smallest measured values of T_{1e}, and a reduction in the thickness of this sample by one half had no significant effect on the value of this parameter. Yet the curve shape and the temperature dependence of T_{1e} established without doubt that a phonon bottleneck was present. It is probable that for this material the parameter L in equation (4.8) was some distance inside the crystal between fast relaxing imperfections and not an external dimension.

F

It is interesting to note the temperature to which the hot phonons corresponded in these two experiments. Faughnan and Strandberg showed that when the spin temperature is infinite (i.e. at the end of the saturating pulse), the phonons in contact with the spins have an effective temperature θ_{ph} given by

$$\theta_{ph} = (1+\sigma)T_b$$

Standley and Wright found a maximum value of $\theta_{ph} = 6\cdot30°$K when T_b was $2\cdot41°$K. Taylor found $\theta_{ph} = 10\cdot5°$K at $T_b = 1\cdot75°$K and $12°$K at $T_b = 3°$K.

A further very convincing experimental proof of the existence of the bottleneck and of the above analysis is afforded in the work of Scott and Jeffries[3] on the paramagnetic ions neodymium and praseodymium in the diamagnetic host lattice lanthanum magnesium nitrate, $La_2Mg_3(NO_3)_{12}, 24H_2O$. The Nd^{3+} ion has a Kramers doublet energy level lowest, split on application of a magnetic field. Measurements were made on two crystals containing 1 and 5 per cent neodymium. The frequency was $9\cdot37$ Gc/s and the relaxation data in the helium temperature region fitted well the equation

$$T_1^{-1} = 1\cdot7T + 6\cdot3 \times 10^9 \exp(-47\cdot6/T)\sec^{-1}$$

That is to say, direct and Orbach terms of the type discussed in Chapter 2 were clearly apparent. Since no phonon bottleneck was observed, it must be concluded that the relaxation rate for the direct process was dominated by the term AT and not by DT^2. Calculated values of the constants A and D for the material in question were $2\cdot6$ and 13 respectively; as shown on p. 65 a phonon bottleneck would not be expected in this case.

The 1 per cent crystal was used in measurements at $34\cdot3$ Gc/s where the relaxation rate was given by

$$T_1^{-1} = 32T^2 + 4 \times 10^9 \exp(-47\cdot6/T)\sec^{-1}$$

Here there is clear evidence of the bottleneck; calculated values of A and D were now 300 and 21 respectively, so in the region of $2°$K, the T^2 term clearly predominates, as observed experimentally. The phonon relaxation time T_{ph} was approximated by the expression

$$T_{ph} = \frac{\text{Crystal half thickness}}{\text{Velocity of sound in crystal}} = \frac{L}{2v}$$

and this approximation appears to be justified by the observed results and their similarity with the calculated data.

In the case of praseodymium, Pr^{3+} is a non-Kramers ion and relaxation was observed in the ground doublet, which was fast relaxing. The occurrence of the bottleneck at low temperatures was therefore not surprising and the experimental data for a specimen containing 1 per cent Pr^{3+} in the double nitrate were well represented by the equation

$$T_1^{-1} = 84T^2 + 4\cdot6 \times 10^{-10} \exp(-54\cdot6/T) + 2\cdot35 T^7 \text{ sec}^{-1}$$

These data refer to a crystal 1·4 mm thick. In order to check the expected concentration and size dependence of the bottleneck rate DT^2, data were taken on a second crystal 2·1 mm thick containing 0·1 per cent Pr^{3+}. The experimental data were now given by

$$T_1^{-1} = 500T^2 + 2\cdot35 \times T^7 \text{ sec}^{-1}$$

over the lower temperature region. The bottleneck factor D was thus increased nearly sixfold. It can be shown that $D \propto (T_{ph}c)^{-1}$, where c is the number of spins per cc, and hence $D \propto (Lc)^{-1}$, where L is the thickness of the crystal (strictly speaking L is the minimum dimension of the crystal). Thus, theoretically, the two values of D should be in the ratio of $(1\cdot4 \times 1)/(2\cdot1 \times 0\cdot1) = 6\cdot7$, in very good agreement with the experimental results in view of the assumption made about T_{ph}.

Brya and Wagner[9] reported observations on trivalent cerium in lanthanum magnesium nitrate, wherein a phonon bottleneck had been observed also by other workers. Their study is of particular interest since it reports what is thought to be the first clear evidence of a phonon bottleneck in the Orbach process. The concentration dependence in the bottlenecked Orbach process has now been checked by Addé et al.[10] for the $\bar{E}(^2E)$ excited state in ruby.

Relaxation measurements involving the phonon bottleneck have been reported by several authors, for example Nash,[11, 12] and Giordmaine and Nash.[13]

References

1. Van Vleck, J. H., *Phys. Rev.*, 1941, **59, 724.**
2. Faughnan, B. W., and Strandberg, M. W. P., *J. Phys. Chem. Solids*, 1961, **19,** 155.
3. Scott, P. L., and Jeffries, C. D., *Phys. Rev.*, 1962, **127,** 32.
4. Stoneham, A. M., *Proc. Phys. Soc.*, 1965, **86,** 1163.
5. Klemens, P. J. *Solid St. Phys.*, 1958, **7,** 1.
6. Eisenstein, J., *Phys. Rev.*, 1951, **84,** 548.
7. Standley, K. J., and Wright, J. K., *Proc. Phys. Soc.*, 1964, **83,** 361.
8. Taylor, P. F., Ph.D. thesis, St Andrews University, 1968.
9. Brya, W. J., and Wagner, P. E., *Phys. Rev.*, 1966, **147,** 239.

10. Addé, R., Geschwind, S. and Walker, L. R. *Proceedings of XV Colloque Ampère*, 1968.
11. Nash, F. R., *Phys. Rev. Letters*, 1961, **7**, 59.
12. Nash, F. R., *Phys. Rev.*, 1965, **138**, A1500.
13. Giordmaine, J. A., and Nash, F. R., *Phys. Rev.*, 1965, **138**, A1510.

Chapter 5

Non-Resonant Measurement
Techniques

The first measurements of spin-lattice and spin–spin relaxation
times were carried out by using non-resonant techniques. An
excellent account of this early work is given by Gorter in his book[1]
and a later summary is given by Cooke.[2] In much of this work the
measurements consist either of dispersion or of absorption
measurements at audio frequencies and the relaxation time is
deduced from an analysis of the data in a manner which will be
described below. There is one obvious and important difference
between such measurements and those referred to elsewhere in
this book. With the resonant techniques, a single pair of energy
levels is studied and the influence of neighbouring and distant
energy levels is apparent through the disturbance of the expected
relaxation between the two levels under study. In the non-resonant
method, magnetic susceptibility data, for example, are taken which
clearly in some way represent an average behaviour over all populated energy levels. Any relaxation parameter thus determined
must itself be an average one and it is not surprising that in certain
cases the observed behaviour cannot be truly explained in terms
of a single relaxation time. There is, moreover, a second difficulty.
Most of the susceptibility measurements are carried out on concentrated paramagnetic samples—indeed the sensitivity of the
method makes it difficult to work with magnetically dilute
materials; in the resonant method we can use samples in which
paramagnetic ions represent less than 1 in 10 000 of the total
metallic ions in the crystal. The theoretical analyses referred to in
Chapters 2, 3 and 4 specifically relate to those cases in which the
paramagnetic ions can be considered as isolated, and it is again not
surprising that much of the non-resonant data is not in very good
accord with these theories. Nevertheless the measurements of the
Leiden and Oxford groups, for example, are important in their own
right, and a brief account of the experimental method and the relevant theory will be given below.

Theory of the method

We shall confine ourselves to those measurements which are made by studying the behaviour of the salt in an oscillating magnetic field, in the simultaneous presence of a relatively large parallel steady magnetic field when spin-lattice effects are to be examined. The frequency of the oscillatory field is variable and relaxation effects are manifested by the appearance of a component of magnetization out of phase with the field, accompanied by a simultaneous diminution in the in-phase component. In the condition so defined we shall clearly measure the differential susceptibility of the salt given by

$$\chi = dM/dH$$

and the existence of in-phase and out-of-phase components of magnetization can be formally represented by writing the susceptibility in the complex form

$$\chi = \chi' - i\chi''$$

The experimental results are then expressible as the variation of χ' and χ'' with the frequency of the oscillating field. The prime task of the theory is to relate these parameters and their variation to the relaxation times.

Cooke has pointed out that the spin–spin and spin-lattice effects can usually be separated. This comes about since, in most salts, spin–spin relaxation times are of the order 10^{-8} to 10^{-10} seconds and so have very little effect on the behaviour of salts examined at frequencies of the order of 10^6 or 10^7 c/s. On the other hand, spin-lattice effects are not detectable in very small fields where magnetization occurs by polarization of the spin energy levels instead of by transfer of spins from higher to lower levels. The reason for this is shown by a simple example. In any paramagnetic material a given spin is subject to a fluctuating magnetic field from the moments appearing at the neighbouring lattice sites. In iron ammonium alum for example, the r.m.s. value of this field is given by Volger, de Vrijer and Gorter[3] as 450 gauss. When an external field is applied, which is very much smaller than the internal field, its principal effect is to change the direction rather than the magnitude of the field on each dipole. The magnetization which occurs comes about by a slight alteration of the orientation of the dipoles, and there is no interchange of energy between them and the lattice. The energy of magnetization is small compared with the energy of interaction of the dipoles and it is taken up by small adjustments of their energy levels. On the other hand, an applied field large compared with the internal field creates a situation where a small increase in the external field changes the magnitude but not the direction of the field on each dipole. The resulting increase in the magnetic moment of the specimen occurs by flips of the dipoles from positions opposing the field to positions in which they assist it, the difference in energy between these two positions being given to the lattice. At intermediate values of the magnetic field both processes contribute to the magnetization, but only the second process, involving change of population of the energy levels, involves spin-lattice relaxation. The two relaxation effects can therefore be studied separately; spin-spin effects are determined by measurements in small steady magnetic fields using high-frequency oscillatory fields, while spin-lattice effects can be determined by measurements in larger magnetic fields at frequencies so low that spin-spin relaxation effects can be neglected.

The following treatment follows that given by Cooke. In treating the effects thermodynamically it is assumed that the relaxation time for the establishment of equilibrium among the spins is so short that the entire spin system is always in thermodynamic equilibrium and so can be treated as a thermodynamic entity separate from the crystal lattice. The spin system can possess a temperature and specific heat, for example, peculiar to its own system, which may differ from those quantities possessed by the lattice. If now a large steady field and a parallel oscillating field are applied to a magnetic salt, the heat of magnetization will cause the temperature of the spin system to oscillate with that field. The oscillations in temperature will be reduced by the thermal contact between the spins and the lattice; clearly, if the period of oscillation of the field is very long compared with the spin-lattice relaxation time, the spin temperature will always be very close to that of the lattice, which is assumed to be constant, and in this case the susceptibility of the salt will be its isothermal value χ_T. The other extreme possibility occurs when the period of the field is so short that there is no appreciable heat transfer between the spins and the lattice; now the susceptibility has a different value which can reasonably be called the adiabatic susceptibility χ_S. We have the following expressions for these two susceptibilities

$$\chi_T = \left(\frac{\partial M}{\partial H}\right)_T \qquad \chi_S = \left(\frac{\partial M}{\partial H}\right)_S$$

where M is the magnetization, H the magnetic field and S the entropy.

Thermodynamic arguments lead to two further equations:

$$\frac{\chi_T}{\chi_S} = \frac{C_H}{C_M} \tag{5.1}$$

and

$$C_H - C_M = T \left(\frac{\partial M}{\partial T}\right)_H^2 \frac{1}{\chi_T} \tag{5.2}$$

In these expressions M is the magnetization in field H, and C_M and C_H are the specific heats of the *spin system* at constant magnetization and constant field respectively. By the use of these equations it is possible to arrive at the specific heats of the spin system, using magnetic data derived during relaxation experiments.

Individual values of χ_T and χ_S give no information on possible relaxation effects. In order to describe these, further information on the heat transfer between the spins and the lattice is needed. In their paper, which is generally regarded as the basis for the non-resonant determination of relaxation times, Casimir and du Pré[4] made the basic assumption that the rate of flow of heat at any instant is directly proportional to the temperature difference between the spin system and the lattice. A more generalized treatment was given by Debye.[5]

We thus write the rate of flow of heat into the spin system as

$$\frac{dQ}{dt} = -\alpha \Delta T \tag{5.3}$$

when the spin temperature exceeds the lattice temperature by ΔT. The constant α in effect measures the strength of the contact between spins and lattice and was called by Gorter the coefficient of thermal contact.

Integrating equation (5.3) gives

$$-\alpha \int \Delta T dt = \Delta Q = C_M \Delta T - T \left(\frac{\partial H}{\partial T}\right)_M \Delta M \tag{5.4}$$

Assuming that the field H varies periodically with a pulsatance ω and that the temperature variation has a similar form, equation (5.4) becomes

$$-\frac{\alpha}{i\omega} \Delta T = C_M \Delta T - T \left(\frac{\partial H}{\partial T}\right)_M \Delta M$$

Now

$$\Delta M = \left(\frac{\partial M}{\partial H}\right)_T \Delta H + \left(\frac{\partial M}{\partial T}\right)_H \Delta T$$

and eliminating ΔT between these two equations, we have

$$\frac{\Delta M}{\Delta T} = \left(\frac{\partial M}{\partial H}\right)_T \frac{C_M + \dfrac{\alpha}{i\omega}}{C_M - T\left(\dfrac{\partial H}{\partial T}\right)_M \left(\dfrac{\partial M}{\partial T}\right)_H + \dfrac{\alpha}{i\omega}} \tag{5.5}$$

This may be written

$$\chi = \chi_T \frac{C_M + \dfrac{\alpha}{i\omega}}{C_H + \dfrac{\alpha}{i\omega}} \tag{5.6}$$

since

$$-\left(\frac{\partial H}{\partial T}\right)_M = \left(\frac{\partial M}{\partial T}\right)_H \qquad \left(\frac{\partial H}{\partial M}\right)_T = \left(\frac{\partial M}{\partial T}\right)_H \cdot \frac{1}{\chi_T}$$

The denominator of (5.6) then follows from equation (5.2). Writing $\chi = \chi' - i\chi''$ equation (5.6) reduces to

$$\chi' = \frac{\chi_T - \chi_S}{1 + \omega^2\tau^2} + \chi_S \tag{5.7a}$$

and

$$\chi'' = (\chi_T - \chi_S)\frac{\omega\tau}{1 + \omega^2\tau^2} \tag{5.7b}$$

where

$$\tau = C_H/\alpha \tag{5.8}$$

Provided that the assumption is valid that the heat flow is proportional to the temperature difference between spins and lattice, equations (5.7) and (5.8) enable the relaxation parameter τ and the adiabatic susceptibility χ_S to be determined either from dispersion measurements which give χ', or from absorption measurements—the heat absorbed in an oscillatory field—which give χ''. If the amplitude of the oscillatory field is H_1, then the heat absorbed per second by 1 gm of the paramagnetic salt is

$$A = \tfrac{1}{2}\omega\chi''H_1^2 \tag{5.9}$$

Fig. 5.1 is taken from Cooke's paper and shows the variation with frequency of χ' and χ'' drawn for the case $\chi_S = 0.2\chi_T$; it also shows the absorption A in arbitrary units.

The Casimir-du Pré theory was applied satisfactorily in the many researches published by the Leiden and Oxford schools, and in the literature the values of τ will be found (in Gorter's book the symbol ρ is used instead of τ). Clearly the forms of equations (5.7) and (5.8) will only be followed when it is a valid approximation to assume that a single relaxation rate covers all the transitions. At very low temperatures this is unlikely to be true as was found to be the case in the experiments, for example, of Kramers, Bijl and Gorter.[6]

If the salt is subjected to a varying field of sufficiently high frequency, it is usually found that there can be detected an additional absorption of heat which can be described by the equation

$$A_S = \tfrac{1}{2}\omega^2 \chi_S \tau_S H_1{}^2$$

Here A_S is the absorption of energy per gram of salt per second and χ_S the adiabatic susceptibility per gram. A_S is proportional to χ_S—

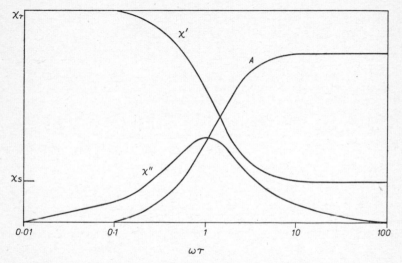

FIG. 5.1. The variation of χ' and χ'' with frequency for $\chi_S = 0.2\chi_T$. Curve A shows the heat absorption per second in arbitrary units (after Cooke[2]).

that is to say, to the susceptibility remaining when spin-lattice energy transfer is excluded—and it is therefore to be ascribed to spin–spin relaxation effects. Since the parameter τ_S has the dimensions of time, it is called the spin–spin relaxation time. It is usually found to be of the order of magnitude 10^{-10} seconds in the concentrated salts used in these measurements, and because it is so small spin–spin effects can only be detected when measurements are made at very high frequencies. Conversely, it generally has very little effect on the spin-lattice relaxation measurements made at much lower frequencies. It is not proposed to discuss τ_S further here, and the reader is referred to the paper by Cooke[2] for details.

It is to be noted that spin–spin relaxation times are usually derived from absolute measurements of the absorption, and not from dispersion.

The above simplified derivation of the Casimir-du Pré equations will suffice to show the probable difference between the parameter τ and the spin-lattice relaxation time T_1 referred to in other sections of this book. As remarked above, by the very nature of the experiment τ is an average over all possible transitions and energy levels, and it will not be surprising if the form of its variation with temperature, for example, differs from that of T_1. The experimental techniques will be discussed briefly below, in order that the difference between these and the resonance-type experiments can be clearly shown.

Experimental methods

(a) *Absorption measurements*: The first experimental detection of paramagnetic relaxation effects by Gorter in 1936 used the absorption technique. As indicated above, if the specimen is in an alternating magnetic field, the energy absorbed per gram of salt per second is given by equation (5.9). If the heat generated in the salt is accurately measured and the frequency and field are known, we have a direct measure of χ''. In the first experiments the frequencies were in the range 10–30 Mc/s and fields at the specimen were of the order of 8 gauss. In later experiments higher fields were used, while in the high-frequency experiments of Volger, de Vrijer and Gorter[3] a $\frac{1}{2}$ kW oscillator gave fields up to 3 gauss at 78 Mc/s. In all absorption experiments the main difficulty is to ensure that unwanted heating is kept to a minimum and that the ambient conditions are maintained constant. To this end the Dutch experiments were usually carried out with the specimens in a glass tube suspended in a vacuum enclosure and connected to a sensitive manometer. The whole was placed in a Dewar vessel containing liquid nitrogen or liquid hydrogen. The tube containing the specimen served both as a calorimeter and as the bulb of a constant volume gas thermometer, by which the rise in temperature could be measured when the oscillating field H_1 was switched on.

The procedure of any experiment was as follows. The specimen was first cooled to the required experimental temperature, and the space surrounding the calorimeter was evacuated, the temperature then being measured at fixed intervals of time to establish the rate of drift. The oscillatory field was then switched on for a few minutes and the rise in temperature was measured. The experiments were made at different temperatures and frequencies and with steady magnetic fields applied parallel or perpendicular to the oscillatory field. It was necessary to convert the measured rates of rise of the gas thermometer pressure into values of χ'' and therefore the heat capacity of the specimen and calorimeter and the amplitude of the alternating field were required. The absolute and accurate determination of these quantities is not easy and often it has been the practice to compare the measurements at different temperatures, frequencies and fields, adjusting the various parameters to give the best fit to the Casimir-du Pré equations. The relaxation times obtained in this way appear to be consistent among themselves while the specific heats so determined are apparently in good agreement with the results of measurements on similar materials by other methods.

(b) *Dispersion measurements*: This method of measurement was first used by Gorter and Brons.[7] In it, the paramagnetic specimen is inserted into an inductance coil which forms part of a radio-frequency tuned circuit, and the resulting change in resonant frequency is measured by a suitable method, usually heterodyne. The frequency change gives the increase in self-inductance of the coil which is proportional to χ', while the out-of-phase component χ'' could in theory be determined from the change in Q of the circuit.

For most paramagnetics measured, the incremental susceptibility of the salt is found to be independent of the frequency of the alternating field, in the absence of an applied steady field. Thus, by taking the susceptibility under these conditions as equal to the static susceptibility χ_T of the salt which can be determined by subsidiary measurements, a calibration of the apparatus is achieved and the results of other experiments can be expressed as the ratio χ'/χ_T; that is to say, one determines the ratio of the increase of inductance at a given steady field to the increase in the absence of the steady field at the same frequency and temperature. When a steady magnetic field is applied parallel to the alternating field, the susceptibility is a function of frequency which diminishes as the frequency is raised, and at very high frequencies it becomes constant again at its adiabatic value χ_S. The form of the variation with frequency follows generally the Casimir-du Pré theory and the experiments can be interpreted to give the spin-lattice relaxation parameter τ and the ratio χ_S/χ_T from which the magnetic specific heat can be determined.

In the Dutch experiments the frequencies varied from 0.1 to 14 Mc/s and the measuring coils generally formed part of the tuned-circuit of an oscillator whose frequency could be compared with a fixed-frequency reference oscillator by mixing the output of the two and detecting the resulting audio difference frequency.

Constant magnetic fields up to 3000 gauss were produced by a large water-cooled solenoid surrounding the Dewar vessel in which the measuring coil was suspended. In this method a change in frequency is interpreted in terms of a change in inductance and it is most important that inserting the paramagnetic salt into the coil should not introduce dielectric changes. These, of course, would result in a change in the self-capacity of the coil and hence in the resonant frequency of the circuit.

(*c*) *Measurements at liquid helium temperatures*: In this method, first devised by Casimir, Bijl and du Pré,[8] the paramagnetic salt is surrounded by a mutual inductance coil system, wherein the primary carries the alternating magnetizing current, and also by a solenoid which gives the required steady field parallel to the alternating one. Changes in the susceptibility of the salt are measured

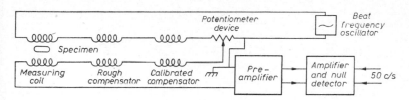

FIG. 5.2. The electrical circuit used by Benzie and Cooke.[9]

by the changes produced in the mutual inductance system. The induced voltage in the secondary coil, due to the component of magnetization of the salt which is in phase with the alternating field, is balanced by a decade type of mutual inductance which has its primary and secondary separately in series with those of the coil surrounding the specimen. The difference ΔM between the value of the compensating inductance M and its zero value in the absence of the specimen gives χ' in some arbitrary units. The secondary voltage due to the out-of-phase component of magnetization is balanced by a potentiometer system which enables the voltage across a resistance in the primary circuit to be injected into the secondary one. If the difference in this resistance from its zero value is ΔR at a frequency $\omega/2\pi$, $\chi'' = \Delta R/\omega$ in the same units as χ'. The arbitrary units are converted into absolute values by using the fact that in zero applied field $\chi'' = 0$ and $\chi' = \chi_T$ whose value is known from other experiments.

The following data are taken from a paper by Benzie and Cooke[9] and serve to illustrate the points just made. The electrical circuit is shown in Fig. 5.2 and illustrates Hartshorn's method described above. The compensating mutual inductance was a four-decade type, graduated in arbitrary units of approximately 0·153 μH. The potentiometer device consisted of twenty-four steps of 0·1 ohm and a 0·1-ohm slide wire shunted to give dividing factors of 10 and 100. The shunts served the double purpose of reducing to a negligible quantity the inductive feed from primary to secondary and increasing the accuracy of reading. The primary circuit could be fed from a stable oscillator giving frequencies of 12·5, 20, 33·3, 75, 175, 500 and 950 c/s, checked against a standard 50 c/s. The

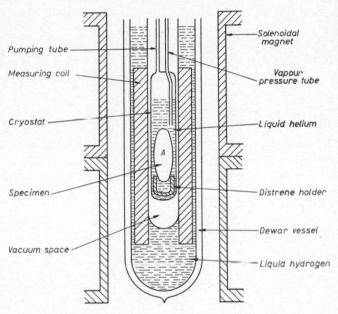

FIG. 5.3. The apparatus used by Benzie and Cooke.[9]

limit of sensitivity of the detector system was 1 μV. The primary alternating current was in the region of 0·1 amp which produced an alternating field of about 4 gauss at the specimen. Parallel steady fields up to 1000 gauss were produced by an additional solenoid.

Fig. 5.3 indicates the form of the apparatus used by these experimenters. Small crystals of the salt used were normally contained in the spheroidal perspex container A, 40 mm long and 10 mm diameter internally. The specimen must have a precise shape when demagnetizing effects have to be taken into account. The experimental procedure was first to measure the susceptibility of the salt at different temperatures. Measurements were made down to approximately 1°K without an external steady magnetic field in order to establish the exact law of variation of the susceptibility of the specimen with temperature; in later parts of the experiment this information could be used to provide a secondary thermometer which could at any time be checked against the vapour pressure thermometer. With the cryostat maintained at a constant tem-

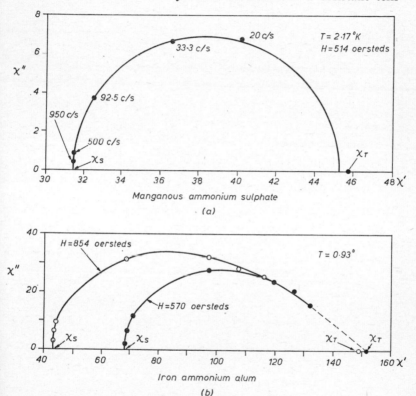

FIG. 5.4. Experimental results for (*a*) manganous ammonium sulphate, (*b*) iron ammonium sulphate (after Benzie and Cooke[9]).

perature by appropriately pumping over the liquid helium, mutual inductance measurements were made at the steady frequencies in zero external fields and in several different values of the steady field. χ' and χ'' were thus determined and, if the results followed the Casimir-du Pré relations, values of relaxation parameters could be obtained. Benzie and Cooke found that only in the case of one salt (manganous ammonium sulphate) did they get a clear indication that the Casimir-du Pré relations were followed at very low temperatures. These results are shown in Fig. 5.4 together with those for iron ammonium alum which clearly indicate a departure from the simple theory. [The form of the equations (5.7a) and (5.7b) is such that a plot of χ' against χ'' should be a semi-circle if a single relaxation time properly describes the system.] These measurements therefore do not necessarily lead to a measurement of the relaxation parameter τ, which alone may not correctly describe relaxation in this spin system.

Experimental values of τ

The experimental results have been summarized and discussed at length by Gorter[1] and, to a lesser extent, by Cooke.[2] We do not propose to discuss them further, and the reader is referred to those publications for details. We reiterate that this step is taken since the concentrated magnetic salts usually employed in the non-resonant experiments do not conform at all to the models used in the theoretical approaches to spin-lattice relaxation, nor are the parameters determined generally directly comparable with those found by the resonance techniques.

References

1. Gorter, C. J., *Paramagnetic Resonance* (Elsevier Publishing Co., 1947).
2. Cooke, A. H., *Repts. on Prog. in Phys.*, 1950, **13**, 276.
3. Volger, J., De Vrijer, F. W., and Gorter, C. J., *Physica*, 1947, **13**, 621, 635 and 653.
4. Casimir, H. B. G., and du Pré, F. K., *Physica*, 1938, **4**, 579.
5. Debye, P., *Phys. Z.*, 1938, **39**, 616.
6. Kramers, H. C., Bijl, D., and Gorter, C. J., *Physica*, 1950, **16**, 65.
7. Gorter, C. J., and Brons, F., *Physica*, 1937, **4**, 579.
8. Casimir, H. B. G., Bijl, D., and du Pré, F. K., *Physica*, 1941, **8**, 449.
9. Benzie, R. J., and Cooke, A. H., *Proc. Phys. Soc.*, 1950, **63**A, 201 and 213.

Chapter 6

Resonant Techniques

The Continuous Saturation Method

Introduction

In the last chapter it was pointed out that one of the very serious drawbacks of the non-resonant method is that the parameter actually measured may not in general be simply related to the spin-lattice relaxation time T_1 defined in Chapter 1; it is more likely instead to be a weighted mean of the relaxation rates of all populated paramagnetic states. With the advent of spin resonance techniques, however, it became possible to measure the actual population difference between a selected pair of energy levels, and this opened up a way of determining the relaxation behaviour related to individual transitions.

The first report of measurements of this type was that of Bloembergen, Purcell and Pound[1] who determined nuclear relaxation times by two different resonant methods. In the first, a continuous saturation method, the authors observed the onset of saturation in the nuclear absorption signal under conditions of increasing incident power, and related this to the spin-lattice and spin–spin relaxation times by means of the Bloch[2] equations. In the second, a transient method, the resonance signal was used to monitor the return of the instantaneous population difference to its thermal equilibrium value after equilibrium had been disturbed by means of a saturating pulse of power. In both methods, the parameter measured was the change in nuclear susceptibility of the material under investigation, this change actually being manifested by a disturbance in the balance of a radio-frequency bridge circuit containing the inductance coil which enclosed the sample. At a frequency in the neighbourhood of 30 Mc/s, H^1 and F^{19} resonances were studied in magnetic fields of about 7000 gauss. Frequencies of 14·5 and 4·5 Mc/s were also used.

The development of the transient method of measurement is described in Chapter 7. In this chapter we discuss the continuous saturation method and its application to electron spin relaxation

87

studies. An early reference to the measurement of relaxation times of paramagnetic ions under resonant conditions is that of Slichter and Purcell[3] who employed microwave pulses from a magnetron to produce saturation in the absorption of microwave energy by a sample located in a resonant cavity. The materials used were manganese and copper sulphates and manganese ammonium sulphate at room temperature and 195°K, and they obtained relaxation times of the order of 10^{-8} seconds.

In 1951 Schneider and England[4] reported the use of this technique at 9·5 KMc/s to study divalent manganese ions in zinc sulphide phosphors. At liquid air temperatures saturation effects were observed with powers of the order of 100 mW and T_1 was estimated to be about 10^{-6} seconds. Further measurements using this technique were reported in 1953 by Eschenfelder and Weidner,[5] and in 1957 by Feher and Scovil,[6] and later by several other workers.

The extraction of numerical values of relaxation parameters from the experimental results requires assumptions to be made, particularly concerning the physical model used to describe the spin system. Several approaches to the problem are given below and the significance of the assumptions made will be brought out.

Theory

The original theoretical treatment of saturation effects, developed by Bloembergen, Purcell and Pound[1] for the nuclear resonance case, specifically applied to a two-level system. Eschenfelder and Weidner[5] extended this theory to include paramagnetic resonances, and Lloyd and Pake[7] generalized the theory to take account of the more usual multilevel system. For simplicity we shall first restrict our attention to a two-level spin system. We have already seen [equation (1.21), p. 12] that, in the absence of a radiation field, the rate at which a disturbed spin system will tend to return to equilibrium with the lattice is given by

$$\frac{dn}{dt} = \frac{n_0 - n}{T_1} \tag{6.1}$$

In the presence of a radiation field however, an extra term must be introduced and the equation becomes

$$\frac{dn}{dt} = \frac{n_0 - n}{T_1} - 2nW_s \tag{6.2}$$

Here W_s is the probability per spin per unit time of a spin transition in the presence of the radio-frequency field, and is assumed to be the same for transitions in either direction between the two spin levels. In the presence of a radiation field which will disturb the equilibrium distribution of the spins between the two levels, the effect of spin-lattice relaxation in depopulating the upper level is opposed by the radiation field which is on balance endeavouring to promote spins from the lower to the upper level. A steady distribution is reached when the two effects are equal in magnitude and the rate of change of population is then zero. Thus equation (6.2) becomes

$$\frac{n_0 - n}{T_1} = 2nW_S$$

and

$$\frac{n}{n_0} = \frac{1}{1 + 2W_S T_1} \tag{6.3}$$

The transition probability W_S for a magnetic dipole transition between two states $|a\rangle$ and $|b\rangle$ is given by

$$W_S = \gamma^2 H_1{}^2 |\langle b | M | a \rangle|^2 g(v) \tag{6.4}$$

where γ is the gyromagnetic ratio, H_1 the amplitude of the radiation field and $g(v)$ the line-shape function. For the case of the two spin levels, $S_z = \pm \frac{1}{2}$, equation (6.4) becomes

$$W_S = \tfrac{1}{4}\gamma^2 H_1{}^2 g(v) \tag{6.5}$$

Substituting this value of W_S back into equation (6.3) gives the steady-state relation

$$\frac{n}{n_0} = \left[1 + \tfrac{1}{2}\gamma^2 H_1{}^2 g(v) T_1\right]^{-1}$$

In equation (6.3), the ratio n/n_0 is the so-called saturation factor S; it is a measure of the amount of saturation produced by the radiation field. If $S = 0$, $n = 0$ and we have complete saturation. It is found convenient to define a quantity $T_2{}^*$ by the equation

$$T_2{}^* = \tfrac{1}{2}g(v)_{max} \tag{6.6}$$

where $g(v)_{max}$ is the maximum value of the line-shape function $g(v)$. Thus $T_2{}^*$ is a measure of the inverse bandwidth and, in the case of a Lorentzian line-shape, $1/T_2{}^*$ is half the difference in frequency between those two points on the resonance curve where the absorption is half the maximum value. (For a homogeneous line $T_2{}^*$ is just the spin–spin relaxation time T_2.) We therefore write the saturation factor S in the form

$$S = \frac{n}{n_0} = (1 + \gamma^2 H_1{}^2 T_1 T_2{}^*)^{-1} = (1 + S')^{-1} \tag{6.7}$$

where $S' = \gamma^2 H_1{}^2 T_1 T_2{}^*$.

It can be shown that this rate equation approach yields the same result for the saturation factor as does the solution of the Bloch equations. From such a solution [equation $(1.3b)$, p. 9] the imaginary part of the complex susceptibility can be derived in the presence of a field $2H_1 e^{i\omega t}$ giving

$$\chi'' = \tfrac{1}{2}\chi_0\omega_0 T_2 \frac{1}{1 + T_2{}^2(\omega_0 - \omega)^2 + \gamma^2 H_1{}^2 T_1 T_2} \tag{6.8}$$

When H_1 is very small, we have the unsaturated case, that is to say

$$\chi''_{\text{unsat}} = \tfrac{1}{2}\chi_0\omega_0 T_2 \frac{1}{1 + T_2{}^2(\omega_0 - \omega)^2} \tag{6.9}$$

And so the saturation factor n/n_0, which is the same as the ratio $\chi''/\chi''_{\text{unsat}}$ for $\omega = \omega_0$, may be written as

$$\frac{n}{n_0} = \frac{1}{1 + \gamma^2 H_1{}^2 T_1 T_2}$$

as before.

A method of measuring T_1 is suggested by the foregoing discussion, namely by examining the absorption signal as a function of increasing power incident upon the specimen. The signal height will be reduced as the incident power is increased, and when its value is one-half of its unsaturated height, the term $\gamma^2 H_1{}^2 T_1 T_2$ is equal to unity. Then a knowledge of H_1 and T_2 [or $g(\nu)$] will be sufficient to enable T_1 to be determined. This is still not an easy task. In order to find $g(\nu)$, the line-shape must be delineated and analysed, and to obtain T_2, the nature of the broadening mechanism must be fully understood. H_1 is the value of the radiation field at the sample and this clearly depends upon several factors, such as the coupling coefficient to the cavity, the geometry of the cavity, its Q-value and the size and location of the sample within it. It would therefore be a lengthy procedure, and one which would need to be followed for every measurement of T_1, to determine fully and accurately values for these parameters. Recently Poole[8] has discussed some of the practical details involved in such an analysis and these are referred to again below.

To overcome most of these objections Eschenfelder and Weidner proposed the following procedure. The cavity system in which the sample is placed is considered equivalent to a series L C R circuit, and an analysis in terms of transmission line theory then leads to an expression for T_1 and T_2 in terms of the reflection coefficients of the cavity on magnetic resonance Γ and off magnetic resonance Γ_0 and the incident microwave power P_i. For spin S it has the form

$$\frac{1-\Gamma}{\Gamma-\Gamma_0} = B\,\frac{T}{g(v)_{\text{max}}} + \frac{4(2S+1)k}{N(hv)^2(1-\Gamma_0)}\,TT_1P_i(1-\Gamma)^2 \quad (6.10)$$

Here B is a constant. The reflection coefficients Γ and Γ_0 may be determined from incident and reflected power measurements using a microwave bridge, and P_i is measured with either a barretter or a thermistor. At very low power levels, Γ becomes almost constant, independent of power, but as the power is increased saturation in the specimen becomes detectable and Γ approaches the value Γ_0. The value of Γ_0 can be determined with the magnetic field well away from the microwave resonance condition.

A plot of the term $(1-\Gamma)/(\Gamma-\Gamma_0)$ against $P_i(1-\Gamma)^2$ theoretically yields a straight line from the slope of which T_1 can be determined. The intercept is proportional to $T/g(v)_{\text{max}}$ and hence T_2 can be found if the line-shape function is known. This procedure enables T_1 to be determined without the necessity of evaluating H_1 or $g(v)_{\text{max}}$.

The multilevel case

In the above analysis for a simple two-level system it is easy to define a unique relaxation time for the return of the spin system to its equilibrium value. Lloyd and Pake[7] pointed out that for a more complicated energy-level scheme a single relaxation time may not readily be associated with each pair of levels. In general, the return to equilibrium from an initially disturbed state will be best described by the sum of several exponential terms of which the contributions will vary throughout the recovery (see for example Chapter 9). Thus, instead of a single rate equation, homogeneous linear equations (k equations for k energy levels) are obtained and the solution of these will describe correctly the recovery of the population of one energy level in terms of transition probabilities connecting each pair of levels.

Lloyd and Pake show that the saturation factor S_{12} defined between levels 1 and 2 of a system involving k levels in all is given by

$$S_{12} = \left(1 + \frac{W_S}{W_{2\to1} + C_{21}^{-1}\sum_{j=3}^{k}W_{2\to j}C_{2j}}\right)^{-1} \quad (6.11)$$

Here W_S defined earlier in this chapter is the stimulated transition probability and $W_{2\to1}$ is the spontaneous transition probability as defined in Chapter 1. The C's are coefficients obtained in solving the above set of equations. For the two-level case this simplifies to

$$S_{12} = \left(1 + \frac{W_S}{W_{2 \to 1}}\right)^{-1}$$

and using the definition

$$W_{2 \to 1} = \frac{1}{2T_1}$$

we return to equation (6.3). Thus equation (6.11) is compatible with our earlier definitions.

Although a system with many levels does not in general permit the definition of a single relaxation time for any pair of its levels, the coefficient of W_S in equation (6.11) is nevertheless a significant quantity in indicating the strength of the relaxation mechanisms influencing any particular transition. For the two-level case this W_S is just half the relaxation rate $1/T_1$. Lloyd and Pake therefore defined the reciprocal of the coefficient of W_r in equation (6.11) as the *relaxation probability* W_r, which is a useful parameter in situations where a relaxation time in the normal sense cannot be uniquely defined. A measurement of the saturation factor will therefore enable W_r to be evaluated, but, unless assumptions are made about the relative importance of the various terms which make up W_r, little information can be derived about transitions between individual pairs of levels, should this be the ultimate aim.

Effects of line broadening mechanisms

The foregoing theory has assumed that the spin system as a whole remains in thermal equilibrium throughout spin resonance. This will only be the case so long as the mechanisms which produce line broadening are of the type defined by Portis[9] as homogeneous, such as dipolar interactions between like spins and interactions with the radiation field. In the case of inhomogeneous broadening, different groups of spins experience different local fields which may arise, for example, owing to spatial inhomogeneity in the applied magnetic field or to dipolar interactions between spins with different Larmor frequencies, hyperfine interactions or crystalline imperfections and so on. These matters are discussed further in §8.1. Here it suffices to note that homogeneous broadening comes from interactions outside the spin system and must be slowly varying over the time required for a spin transition. Thus the resonance line is imagined to be composed of many overlapping resonance lines, each homogeneous and each of width approximately equal to $1/T_2$. It is the envelope of these spin packets which is experimentally observed. Because of this nature of the resonance line, there is a very important distinction to be drawn between the saturation behaviour in the two cases. On saturation of a homogeneous Gaussian line, there is an appreciable broadening and it is the central portion which appears to be flattened. The individual packets constituting an inhomogeneous line also behave in this way, but the width of the envelope is then governed entirely by the distribution of the local fields and hence of the spin packets. Consequently, a broadening of the component packets may have negligible effect on the width of the envelope, and the flattening of the entire line only is observed, the shape remaining the same at all stages.

Portis has calculated the form of the saturation factor S for the two cases and has investigated experimentally how it depends upon the intensity of the applied radiation field for the case of F centres in KCl. The observed line is Gaussian in shape and the linewidth arises from hyperfine interactions. There is very poor agreement between the experimentally observed saturation behaviour and the simple theoretical predictions based upon a homogeneous line, but the results fitted extremely well when the theory was modified on the assumption of an inhomogeneously broadened line of which the spin packets were Lorentzian in shape. Certain basic assumptions were made about the frequency of the modulation envelope of the microwave field and about the value of T_2 resulting in an estimate of $T_1 = 2 \cdot 5 \times 10^{-5}$ sec at room temperature.

A further complication which may arise in the case of inhomogeneous lines derives from the fact that radiation of one frequency applied to the sample will be absorbed only by the spin packet which is precisely resonant with it. Communication of this energy to other spin packets may be a slow process and, if it is slower than the spin-lattice relaxation rate of the resonant packet itself, this latter will behave almost independently of the rest and negligible saturation of the whole line will be observed. This accounts for the well-known hole burning experiments of Bloembergen, Purcell and Pound[1] in which radiation applied to one portion of a resonance line appeared to produce a hole in the line which rapidly filled on removal of the saturating source.

Experimental details

It is appropriate now to examine some of the ways in which saturation data have been collected and the various approaches to the problem of extracting a value of T_1 from them. Basically, little modification need be made to a conventional spectrometer in order to carry out these measurements, the facilities required depending upon the approach to the analysis. That of Eschenfelder and Weidner for example requires the measurement of the reflection coefficient when looking into the cavity arm and the measurement of the incident microwave power. These are standard microwave measurements and the reader is referred to books such as that by Barlow and Cullen[10] for details. A critical assessment of the validity of these methods has been made (Vaughan[11]), and values of T_1 calculated from some of them have been compared with pulse saturation measurements (Taylor[12]) for the case of the material $[(C_4H_9)_4N]_2$ $Cu[S_2C_2(CN)_2]_2$, with varying degrees of success. The conclusions drawn from this work will be discussed later.

(*a*) The most obvious method of obtaining a value for T_1, as mentioned earlier, is to observe the amplitude of the absorption signal (for example as displayed in an oscilloscope at the output of the spectrometer) as a function of incident power and to calculate T_1 directly from equation (6.7). Assuming the validity of this equation, no additional equipment is required, apart from incorporating into a conventional spectrometer system some means of measuring the incident microwave power. Certain other assumptions need to be made, however, these being that (i) the detection system is linear over the range of power levels employed, (ii) there is a known functional relationship between the incident power P_i and the radiation field at the sample H_1, and (iii) a value for T_2 is known.

A slight variation of this, obviating the need to measure the power level *in situ*, requires two accurately calibrated attenuators, one placed in the source arm and the other in the detector arm of a microwave magic-tee bridge. The attenuators are varied together so as to keep the total attenuation, and hence the detector crystal current, constant, and any deviation from linearity in absorption of power may be detected. This method was used by Kipling *et al.*[13]

(*b*) If some form of the resonant absorption line-shape can be assumed, for example a Lorentzian form;

$$f(H) = \frac{y_m}{1 + \left(\dfrac{H - H_0}{\frac{1}{2}\Delta H_{\frac{1}{2}}}\right)^2} \tag{6.12}$$

it can be shown (see, for example, Poole[8]) that, in terms of the peak absorption height y_m and peak-to-peak derivative height $y_m{}'$, the saturation factor S is given by

$$S = \left[\frac{y_m/H_1}{\underset{H_1 \to 0}{\text{Lim}} (y_m/H_1)} \right] = \left[\frac{y_m'/H_1}{\underset{H_1 \to 0}{\text{Lim}} (y_m'/H_1)} \right]^{3\,2} \quad (6.13)$$

where $\underset{H_1 \to 0}{\text{Lim}} (y_m/H_1)$ refers to the value of this ratio for very small values of H_1, that is to say, when the degree of saturation is negligibly small.

A series of E.S.R. spectra would be recorded, with microwave power varying from a condition of negligible saturation [corresponding to $S' \ll 1$, equation (6.7)] to one of pronounced saturation ($S' \gg 1$), and T_2 and T_1 calculated as follows:

$$T_2 = \frac{2}{\gamma \Delta H_{\frac{1}{2}}{}^0} = \frac{2}{3^{\frac{1}{2}} \gamma \Delta H_{pp}{}^0} = \frac{1\cdot3131 \times 10^{-7}}{g \Delta H_{pp}{}^0} \text{ sec} \quad (6.14)$$

where $\Delta H_{\frac{1}{2}}{}^0$ is the width of the line at half power points under conditions of negligible saturation, and $\Delta H_{pp}{}^0$ is the peak-to-peak derivative linewidth under the same conditions, and

$$T_1 = \frac{3^{\frac{1}{2}} \Delta H_{pp}{}^0}{2} \left[\frac{(1/S) - 1}{H_1{}^2} \right] = \frac{0\cdot985 \times 10^{-7} \Delta H_{pp}{}^0}{g} \frac{S'}{H_1{}^2} \text{ sec} \quad (6.15)$$

where both $\Delta H_{pp}{}^0$ and H_1 are in gauss.

Poole presents a sample calculation using as input data the linewidth ΔH_{pp} and derivative height y_m' for varying values of input power P_i. He shows that the useful range of saturation is limited by the fact that at low powers S differs very little from unity, so that the factor $(1/S - 1)$ is very small, and at high power levels y_m' is too small to measure accurately.

(c) By putting $dy_m/dH_1 = 0$, or $dy_m'/dH_1 = 0$ in the theory of (b), it is possible to obtain a value for T_1 from the maximum value of either y_m or y_m' as a function of power, in which case, from equation (6.12)

$$T_1 = 5\cdot7 \times 10^{-8} \Delta H_{\frac{1}{2}}{}^0/gH_1{}^2 \text{ sec from maximum in } y_m$$

or $\quad T_1 = 4\cdot99 \times 10^{-8} \Delta H_{pp}{}^0/gH_1{}^2 \text{ sec from maximum in } y_m'$

In this way, instead of having to calculate values of T_1 for all values of S, as in (b) above, one calculation only need be made using the maximum value of say y_m' obtained by plotting y_m' against microwave attenuation (or power level). y_m' can be measured directly from the chart recorder display of a conventional spectrometer.

(*d*) All the above methods require an estimation of H_1 at the sample. Singer and Kommandeur,[14] however, used a method in which they monitored the absorption signal from a ruby sample located in the cavity, this being used as a standard against which to normalize the amplitude of another sample absorption. They assumed that there were no saturation effects with the ruby; the

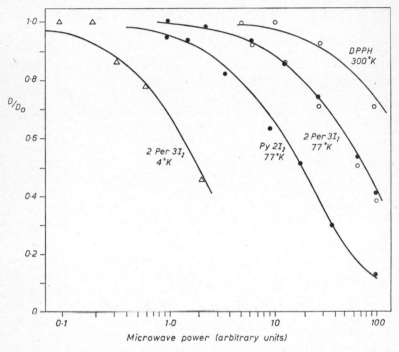

FIG. 6.1. Saturation curves for the pyrene and perylene complexes at low temperatures (Singer and Kommandeur[14]).

absorption D_0 of the ruby at any given power level was then strictly proportional to H_1, and they therefore merely needed to investigate the ratio D/D_0 with incident power, D being the absorption due to the sample under investigation. Any variation of D/D_0 from unity therefore indicates some degree of saturation (see Fig. 6.1).

There is now experimental proof that ruby saturates at relatively low power levels, unless the Cr^{3+} concentration in the ruby is high (see Chapter 9). It is vital that one ensures that the reference sample chosen does not saturate under the conditions of the experiment.

Singer and Kommandeur overcame the difficulty of making an absolute measurement of T_1 by comparing the saturation behaviour of the sample under test with that previously observed using a sample of DPPH (see Fig. 6.1). Lloyd and Pake,[15] and Bloembergen and Wang[16] found that at room temperature, for DPPH, T_1 is very nearly equal to T_2 and is of the order 6×10^{-9} sec. Hence, if P_{DPPH} and P_{sample} refer to the incident power levels required for the same degree of saturation (for example when D/D_0 is 0·5, then $T_1 T_2 = (P_{DPPH}/P_{sample}) T_1{}^2{}_{DPPH}$. T_2 is still required and must be calculated from linewidth measurements. Care must be taken to ensure that the line is homogeneously broadened so that T_2 is a meaningful parameter. This can be assured by observing that the line broadens as it saturates, as indicated above.

(*e*) The method of Eschenfelder and Weidner[5] (see p. 90) obviates the necessity for knowing either H_1 or T_2. This method, however, seems to have very limited application. At very low-power levels, of the order of a few microwatts, it is exceedingly difficult to obtain meaningful values of Γ, and so the range of values of T_1 measurable by this technique is accordingly restricted. Also, as Eschenfelder and Weidner themselves remark, only a relatively small number of their measurements produced the required straight-line graphs, all other results being meaningless.

(*f*) When investigating the saturation behaviour using a method in which modulation is employed, the frequency of modulation of the radiation field may be an important factor in determining the type of behaviour observed. In the above theory the power level has been treated as varying very slowly compared with T_1, so that the spin system was considered to be in a steady state. If, however, the modulation frequency ω_m is of the order of T_1 or greater, the saturation behaviour may be altered. If $\omega_m T_1 < 1$, the spin system or the individual spin packets in the case of an inhomogeneous line) may follow the variations in microwave power and the saturation factor will then also be periodic at the frequency ω_m. If, however, $\omega_m T_1 > 1$, the spin system will not follow variation in power but will react to an average applied magnetic field and will appear to saturate at this average power level. The modulation frequency at which this change in behaviour is observed may therefore be used to estimate T_1.

Such frequency modulation methods have been reported by Hervé and Pescia,[17] for example. Their theory is based on the Bloch equations and is thus subject to the criticisms discussed above but is more easy to interpret because of the lack of any necessity of knowing H_1 or T_2. The only requirement is frequency measurement but the large number of observations required for any single determination of T_1 makes it a rather tedious process. One advantage seems to be the ability to measure fast relaxation times, less than 10^{-7} sec, and such results for some manganese salts have been reported by Pescia and Hervé.[18] Zueco and Pescia[19] later simplified this method by modifying the theory so that only a few measurements were required to calculate T_1.

Carruthers and Rumin[20] used a combination of the continuous saturation method and the non-resonant method. They observed the change in the dispersion at audio modulation frequencies when the sample was subject to varying degrees of saturation produced by a resonant microwave signal.

Assessment of the methods

The main advantage that the continuous saturation method has over some of the more direct methods is the lack of any special equipment required. A further point in its favour is that it may be possible to apply this method to different systems having a very wide range of T_1 values, subject to sufficient power being available to saturate a system having a short relaxation time at one end of the scale, and to sufficient sensitivity being available to detect the very weak absorptions resulting from using very low incident power levels at the other end of the scale.

The main disadvantage arises from the fact that this method is inherently an indirect one. It relies on an assumption of the physical model of the magnetic ion and on the validity of the Bloch equations.

The assessment so far has been theoretical. It appeared to the authors that no experimental data were available which reliably compared values of T_1 obtained both by continuous saturation and pulse saturation techniques. Such a study was therefore made[11] on single crystals of the material di(tetrabutylammonium) bis(maleonitrile dithialato)CuII, referred to from here on as Bu_2CuMNT, for which Taylor[12] had published pulse-saturation results. Although the following comments relate specifically to the measurements on this material, the broad conclusions apply quite generally. It is unfortunate that many papers fail to mention the assumptions made in the analysis of the results, and some fail even to record the approach used, yet, as we will show, these factors may influence markedly the numerical value of T_1 calculated from such saturation data.

Approach (a), above, gave by far the most consistent set of results for the material investigated. Since the relaxation times under discussion were rather long for electron spins (of the order of 1 sec), power levels of the order of 0·1 μW were required to observe the onset of the saturation effects at 4·2°K and care had to be taken to ensure the linearity of the detection system at such low power levels, the noise level forming an appreciable part of the detected signal.

The assumption of the line-shape function in approach (b) limits its applicability, since this function may not always be amenable to a mathematical description. One very rarely observes a truly Lorentzian, homogeneous line, so it is not surprising that the results calculated using the formulae derived by Poole[8] should be at times at variance with those calculated in other ways; indeed, they were found also to be inconsistent amongst themselves at different power levels. The determination of a line-shape function each time a value is to be calculated would be a formidable task.

TABLE 6.1. Relaxation times measured by different methods at $T = 4\cdot2$°K. The pulse-saturation technique gave a value of $T_1 = 583$ msec at this temperature.[11]

Method*	T_1 (msec)	P_i (μW)	Method*	T_1 (msec)	P_i (μW)
b	450	1000	a	897	15
b	425	100	c	350	20
b	575	10	d	200	50
b	980	1	e	35	100–400
b	1350	0·1	e	100	10–100
			e	450	1–10

* Methods listed alphabetically as in text.

The same criticisms hold also for methods (c) and (d) above, since the difficulty arises in relating the saturation behaviour of the derivative curve to that of the absorption line. How far down the *derivative* saturation curve must one go in order to obtain a value of $S' = 1$?

For a *given degree* of saturation, all three methods gave results which were well within an order of magnitude of that obtained by the pulse-saturation method (see Table 6.1) and which exhibited the same temperature dependence, but their lack of consistency and their variation with power level make it impossible to have confidence in any single determination. Results obtained using method (e) produced poor straight-line graphs, one difficulty encountered here being the purely practical one of measuring standing wave ratios at very low-power levels (10^{-5} to 10^{-7} W).

Except in certain cases, it is difficult to obtain an exact relation between the microwave power incident on the resonant cavity P_i and the magnitude of H_1, the radiation field *at the sample*. It is usually concluded that

$$H_1{}^2 = KP_i$$

but different authors often quote widely varying values for K. If P_i is in watts, and H_1 in gauss, $10 > K > 0.1$, depending, of course, on the cavity configuration. A mean value of $K \sim 1$ is not unrealistic, but an uncertainty of a factor of 3 or 4 in the resulting value of T_1 must be accepted.

The width of the resonance line in Bu_2CuMNT, which was neither exactly Lorentzian nor Gaussian in shape, was about 75 gauss, yet split (at other orientations) into nine component lines each about 40 gauss wide. It was not, therefore, possible to equate the linewidth to $1/(\gamma T_2)$, even though some slight degree of saturation broadening was observed. There is thus an additional uncertainty in the absolute value calculated for T_1, a factor of 5 or 6, introduced by assuming a value for T_2. The agreement between the results calculated in these various ways (shown in Table 6.1) is therefore remarkably good in the circumstances. It would appear that order-of-magnitude results for T_1 are the best that can be expected using the saturation technique in all but exceptional circumstances.

The phonon bottleneck in continuous saturation measurements

In Chapter 4 we saw how the rate equations for the disturbed spin system were modified in the presence of a phonon bottleneck. In a completely analogous manner, the rate equation (6.2) describing the behaviour of the spin system in the presence of a radio-frequency field can also be modified (Jeffries[21]). The new rate equation is obtained by adding a term $(-2nW_S)$ to equation (4.2) when we obtain, using the nomenclature of Chapter 4,

$$\frac{dn}{dt} = -T_1(\mathrm{d})^{-1} \tanh\left(\frac{\delta}{2kT}\right) [n(2p+1)-N] - 2nW_S \quad (6.16)$$

The steady-state solution of this equation is now

$$\frac{n_0}{n} = S'\sigma[\tfrac{1}{2}\{(1-S') + [(1+S')^2 + 4\sigma S']^{\frac{1}{2}}\} - 1]^{-1} \quad (6.17)$$

which simplifies to the more usual equation (6.7) when $\sigma \ll 1$. σ is the bottleneck factor, which is effectively the ratio of the energy exchange rate between spins and phonons to that between phonons and bath. The saturation behaviour of a line in the presence of a bottleneck is therefore expected to be considerably different from the usual situation.

Such saturation behaviour has been observed for the Cu salt (discussed above) for which a T^{-2} temperature dependence, indicative of phonon bottleneck, had been obtained. Fig. 6.2 shows values of $S = n/n_0$ calculated from equations (6.7), (6.13) and (6.17), together with experimentally measured saturation curves

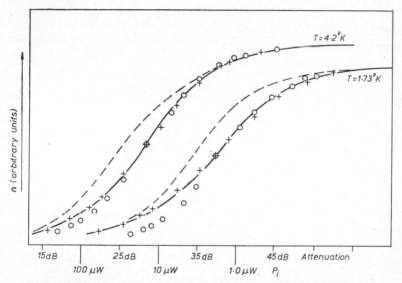

FIG. 6.2. Typical saturation curves at $4\cdot2°$K and $1\cdot73°$K. (n in arbitrary units, Vaughan.[11]) Continuous curves show experimental saturation results obtained using method a. Broken curves indicate results calculated from method b. ○ indicates values of saturation parameter calculated from equation (6.7). + indicates values of saturation parameter calculated from equation (6.17). Both of these are normalized to the experimental values of T_1 at the points ⊕.

for this material (suitably normalized). The best fit is obtained with equation (6.17), using values of σ obtained from pulse-saturation measurements. This fit is quite sensitive to the value of σ.

From equation (6.7), repeated here for reference:

$$S = \frac{1}{1 + S'} = \frac{1}{1 + \gamma^2 H_1^2 T_1 T_2} \qquad (6.7)$$

T_1 can be calculated from the saturation curve by setting $S' = 1$ at the point where S becomes $\frac{1}{2}$. If, however, one makes $S' = 1$

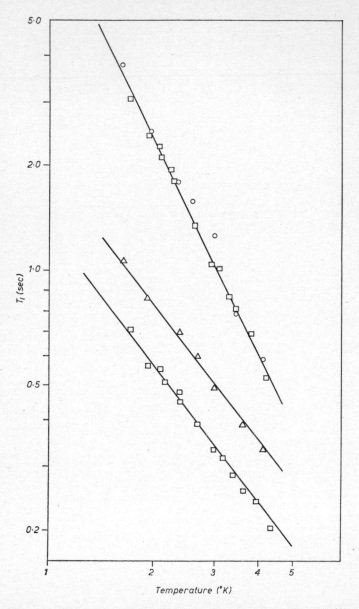

FIG. 6.3. Temperature dependence of relaxation times (Vaughan[11]). In the top curve □ indicates pulse-saturation measurements (Taylor[12]) and ○ continuous saturation measurements, method a, normalized to T_1 = 583 msec at $4.2°$ K. In the middle curve $\triangle$ refers to values of T_1' calculated from pulse-saturation results using equation (6.17) and $\sigma = 1.8$. In the bottom curve □ indicates values of T_1' calculated from pulse-saturation results using equation (4.8) and $\sigma = 1.8$.

in equation (6.17), the value obtained for S will no longer be $\frac{1}{2}$ but is a complicated function of σ. For example, it ranges from 0·372 when $\sigma = 1\cdot8$ to 0·298 when $\sigma = 5\cdot0$. Values of T_1 calculated in this way were found to exhibit a temperature dependence more in the region of T^{-1} than T^{-2}. In fact the same temperature dependence was found as was obtained using the normal bottleneck theory (equation 4.8) on the pulse-saturation results (Fig. 6.3). The latter values of T_1 correspond to the true direct relaxation time $T_1(d)$ of Chapter 4, whereas the former correspond to $T_1(e)$.

As long ago as 1958 Strandberg[22] observed that, if phonon relaxation effects are significant, then these may influence paramagnetic saturation measurements through their dependence on the saturation parameter, and that, in order to separate phonon relaxation effects from spin-lattice relaxation effects, the saturation parameter must be observed and analysed over a wide range of saturation conditions. Subsequent theoretical and experimental results therefore seem to confirm this observation.

References

1. Bloembergen, N., Purcell, E. M. and Pound, R. V., *Phys. Rev.*, 1948 **73,** 679.
2. Bloch, F., *Phys. Rev.*, 1946, **70,** 460.
3. Slichter, C. P., and Purcell, E. M., *Phys. Rev.*, 1949, **76,** 466.
4. Schneider, E. E., and England, T. S., *Physica*, 1951, **17,** 221.
5. Eschenfelder, A. H., and Weidner, R. T., *Phys. Rev.*, 1953, **92,** 869.
6. Feher, G., and Scovil, H. E. D., *Phys. Rev.*, 1957, **105,** 760.
7. Lloyd, J. P., and Pake, G. E., *Phys. Rev.*, 1954, **94,** 579.
8. Poole, C. P., *Electron Spin Resonance* (John Wiley Interscience, 1967).
9. Portis, A. M., *Phys. Rev.*, 1953, **91,** 1071.
10. Barlow, H. M., and Cullen, A. L., *Microwave Measurements* (Constable Co. Ltd, 1950).
11. Vaughan, R. A. To be published.
12. Taylor, P. F., Ph.D. thesis, St Andrews University, 1968.
13. Kipling, A. L., Smith, P. W., Vanier, J., and Woonton, G. A., *Can. J. Phys.*, 1961, **39,** 1859.
14. Singer, L. S., and Kommandeur, J., *J. Chem. Phys.*, 1961, **34,** 133.
15. Lloyd, J. P., and Pake, G. E., *Phys. Rev.*, 1953, **92,** 1576.
16. Bloembergen, N., and Wang, S., *Phys. Rev.*, 1954, **93,** 72.
17. Hervé, J., and Pescia, J., *Proceedings of 12th Colloque Ampère*, 1963, 335.
18. Pescia, J., and Hervé, J., *ibid.*, 347.
19. Zueco, E., and Pescia, J., *Comptes Rendu*, 1965, **260,** 3605.
20. Carruthers, J. A., and Rumin, N. C., *Can. J. Phys.*, 1965, 43, 576.
21. Jeffries, C. D., *Dynamic Nuclear Orientation* (John Wiley Interscience, 1963).
22. Strandberg, M. P. W., *Phys. Rev.*, 1958, **110,** 65.

Chapter 7

Techniques Employing Microwave Pulse Saturation with Monitored Recovery

§7.1

USE OF E.S.R. AS THE MONITOR

It was seen in Chapter 1 that, for the simple case considered, the rate of change of the population difference between two energy levels in a paramagnetic ion after the removal of the disturbing signal is given by

$$\frac{dn}{dt} = \frac{n_0 - n}{T_1}$$

where the spin-lattice relaxation time T_1 is defined by this equation. The solution takes the form

$$n = n_0[1 - \exp(-t/T_1)]$$

which shows that, after the disturbance, the rate at which the spin system returns to its thermal equilibrium distribution follows an exponential law, characterized by the time constant T_1. This indicates that T_1 may be derived from measurements of the variation of population difference as a function of time. Electron spin resonance admirably suits this purpose, since the amplitude of the absorption signal itself is readily shown to be directly proportional to the instantaneous difference in population between the two levels concerned.

In an E.S.R. experiment the paramagnetic sample is usually placed in the resonant cavity of the spectrometer. The quality factor Q is defined as

$$Q = \frac{\omega \times \text{Energy stored}}{\text{Average power dissipated}}$$

Power will be dissipated both in the cavity and in the paramagnetic sample, and Q-factors, Q_0 and Q_m respectively, may be defined corresponding to each of these loss media. From the above equation we may write

$$\frac{1}{Q_m} = \frac{P_m}{\frac{\omega}{8\pi}\int_c H_1{}^2 \, dV} = \frac{P_m}{\frac{\omega}{8\pi} \bar{H}_1{}^2 V_c} \tag{7.1}$$

where P_m is the power absorbed by the sample in the presence of the radio-frequency magnetic field H_1 and $V_c = (\int_c H_1{}^2 dV)/\bar{H}_1{}^2$ and is the effective volume of the cavity. $\bar{H}_1{}^2$ is a weighted average value of the microwave field in the cavity. It follows that, provided the other parameters in equation (7.1) remain constant, $1/Q_m$ is directly proportional to the power absorbed by the sample.

$$\frac{1}{Q_m} \propto P_m \tag{7.2}$$

If the probability per second for a spin to make a transition in the presence of the radio-frequency field is W_S, then on balance in unit time $W_S n$ quanta of energy will be absorbed from this radio-frequency field and

$$P_m = W_S n h\nu \tag{7.3}$$

Hence from equations (7.2) and (7.3) we have

$$\frac{1}{Q_m} \propto n$$

Thus the losses due to the presence of the sample in the cavity are a measure of n, the difference in the populations of the two levels involved. We must now show that the quantity which is detected by the spectrometer is proportional to this power loss.

According to Feher,[1] the change in voltage reflected from the cavity ΔV_r is directly proportional to the imaginary part of the complex susceptibility of the sample χ''. Moreover, if the paramagnetic losses are only a small proportion of the total losses in the cavity and if the other losses remain constant, the change in the loaded Q of the cavity is also proportional to χ'' (see, for example, Assenheim[2]). It is usual for the detector output of a spectrometer to be proportional to the input voltage and it therefore follows that the detected signal in these circumstances is a measure of the instantaneous difference in population between the two levels concerned.

In carrying out the experiment, the specimen is placed in the resonant cavity of the E.S.R. spectrometer, and the power and sensitivity are so arranged that the resonant absorption may be observed without appreciable disturbance of the spin system by the microwave power. The magnetic field is adjusted for resonance and the sample then subjected to a pulse of microwave power at the resonant frequency. As soon as the pulse has ended, the low-power E.S.R. signal is observed, and the behaviour of the amplitude of this signal as a function of time indicates the manner in which the population difference returns to its thermal equilibrium value. If a linear detection system is used, the relaxation time constant may be obtained from a display of the recovery curve, in a manner to be described below.

Basic requirements of the method

The apparatus is shown in the outline block diagram of Fig. 7.1. There, P is a source of microwave pulses at the same frequency as the low-power monitor source M. The pulse width and repetition frequency will usually be variable and it may be that the frequency

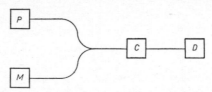

Fig. 7.1. Block diagram of the basic experimental requirement for a pulse-saturation spectrometer.

of both P and M can also be changed. C is a microwave system containing the sample and is usually a resonant cavity, while D is some form of microwave detection system. In practice several ways have been devised for obtaining the two sources and there are various forms of detection system. We shall first discuss some of these components separately and then give specific examples of typical spectrometers which have been used.

Microwave sources

One widely used technique is to employ two separate microwave signals, one from a low-power source of variable amplitude tuned to the resonant frequency of the cavity, and the other from a high-power source (e.g. klystron or magnetron) which supplies the saturating pulses. These signals are brought together, for example, through a directive feed system and are thence fed together to the cavity. Ideally, the two sources must both oscillate at exactly the same frequency so that the maximum power from the pulse source is coupled to the sample and the monitor signal corresponds exactly to the peak of the absorption signal. In practice it is generally sufficient to tune the low-power signal exactly to the frequency of the resonant cavity and exactly to the centre of the absorption line, while the high-power pulse frequency is required only to fall within the magnetic resonance linewidth. An example will perhaps make this last point clear. At 10 Gc/s, if the resonance occurs at about 3000 gauss corresponding roughly to $g = 2$, a linewidth of, say, 20 gauss, measured at fixed frequency and with variable field, corresponds to a linewidth of nearly 60 Mc/s. If this line is homogeneously broadened the energy supplied to one part of it is distributed throughout the whole line in a time of the order 10^{-8} sec, very rapidly compared with the duration of the pulse. Hence, so long as we can put into this resonance line enough power in a pulse, even as much as 200 Mc/s away from the resonant frequency, we are still within the wings of the absorption line and the whole line will be saturated. The same argument applies to an inhomogeneously-broadened line provided that the spin diffusion time between individual spin packets (see p. 148) is short compared with the saturating pulse duration and also with T_1. If this is not the case, there may be complicating effects which make difficult the measurement of T_1 but which also make possible the measurement of spin diffusion times as will be shown below (p. 121).

The second possibility is to derive both signals from one high-power source, and several methods have been used to this end. Scott and Jeffries[3] and Standley and Wright[4] both used a high-power CW klystron, utilizing its full power to saturate the system and interposing a high attenuation path when a low-power signal was required for monitoring purposes. The attenuators in both cases made use of semiconducting diode switches (Feldman and McAvoy,[5]) suitably mounted across the waveguide. The impedance offered by these diodes depends strongly on the d.c. bias applied to them. With a reverse bias of a few volts, the attenuation may be as much as 25 dB, whereas with a forward bias this is reduced to an effective insertion loss of only about 1 dB. If two such switching elements are placed an odd number of quarter guide-wavelengths apart, the attenuation produced by one will add to that of the other in both conditions; hence, far greater isolation can be produced, but at the expense of greater insertion loss.

In order to obtain high isolation without an accompanying high insertion loss, Standley and Wright used a microwave bucking circuit to enhance the attentuation, as shown in Fig. 7.2. By this device, monitor power levels as low as 10^{-9} W were achieved.

Voltage pulses from a conventional pulse generator may be used to activate these switches and a fall-time of the microwave pulse of a few microseconds was easily obtained by Standley and Wright. The limiting factor was the shape of the voltage pulse at the diode switch, and careful attention to the loading of the pulse generator should enable shorter fall times to be achieved.

A simpler alternative method employing a single source is to use a low-power CW klystron and to pulse a travelling-wave tube amplifier in the waveguide run in order to produce the saturating

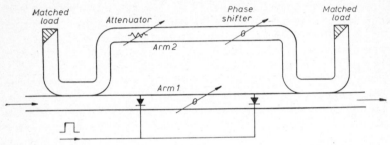

FIG. 7.2. Microwave bucking circuit to enhance the switching action of the diodes.

pulses. This method was first used by Bowers and Mims[6] and later by Lees.[7] The obvious advantage of using a single microwave source is that both signals are of necessity at the same frequency.

Good frequency stability is most desirable in the monitor signal, since relatively small amounts of frequency modulation on that signal are converted to amplitude modulation by the shape of the cavity resonance and this may well mask the signal produced by the sample. In the X-band system used by the authors (see p. 109) the short-term stability of better than one part in 10^6 was more than adequate.

Amounts of pulse power required to saturate a spin system will vary from a few milliwatts to many watts, depending on the relaxation time of the paramagnetic ions under consideration. A working rule to which there may be many exceptions is that a power of 1 mW CW incident on the cavity will just begin to saturate a spin system having a relaxation time of 1 msec. If T_1 were increased to 1 sec, for example, the monitor power used would need to be less than 1 μW in order that some degree of saturation of the spin system might be avoided. For this reason power levels between

10^{-5} and 10^{-8} W are often necessary when monitoring the recovery, and these low powers present some difficulties in detection which are discussed in the next section.

Detectors

When short relaxation times are being measured, of the order of microseconds, relatively high monitor powers, tens of milliwatts at least, can be used, and the detection of these then presents little problem. One may use simple crystal detection techniques followed by wide-band amplifiers, provided that there is sufficient signal intensity under these conditions, and the signal may then be fed directly to an oscilloscope where the recovery can be displayed. Unfortunately these detection conditions rarely obtain and most relaxation time measurements have been made at low temperatures where T_1 is often longer than 1 msec, with monitor powers correspondingly reduced to 1 μW or less. The technique of detection then involves the superheterodyne principle (e.g. Wilmshurst[8]). The usual arrangement is to employ a second klystron operating at a frequency different from that of the monitor, and to mix these signals at the input to the amplifier. The difference signal is taken as the intermediate frequency (i.f.) and the i.f. amplifier must have sufficient bandwidth to pass the required signal undistorted even if the frequencies of the two klystrons should drift very slightly. It is possible for an i.f. bandwidth to be as great as 14 Mc/s, with a centre frequency of 30 or 45 Mc/s. Such a bandwidth ensures that the frequency locking problem is not severe; that problem becomes much more important when an i.f. bandwidth of only 3 Mc/s is used. The requirement of frequency stability is not difficult to satisfy at X-band where it is often possible to use a free-running klystron as a local oscillator if one uses a wide-band i.f. amplifier, but at higher microwave frequencies the stability problem is serious. At Q-band for example, the stability becomes one of ± 7 Mc/s in 35 000, i.e. it becomes a requirement for stability in frequency of better than one part in 5000. Brown *et al.*[9] have reported a way of overcoming this problem by deriving both monitor and local oscillator signals from one klystron, the bandwidth required in the amplifier system now being limited not by the relative frequency stabilities of the signals but by the highest frequency component in the recovery curves.

A typical experimental arrangement

In order to illustrate the facilities required in practice by a pulse-saturation equipment the system used at present by the authors will be described. This comprises a complete superheterodyne spectrometer, together with the pulse-saturation equipment, so that it is possible to measure a spectrum and then immediately to carry out the relaxation measurements.

The equipment is based upon a Decca X3 spectrometer which provides a phase-locked microwave source at 9270 Mc/s, and a

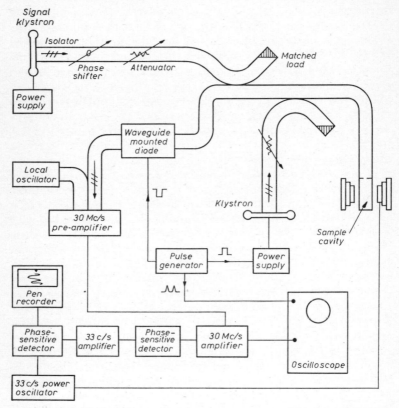

Fig. 7.3. Block diagram of spectrometer system used by the authors based on a Decca X3 spectrometer.

similarly phase-locked local oscillator with phase-sensitive detection both at 30 Mc/s and at 33 c/s. In order to allow the apparatus to be used as pulse-saturation equipment, modifications were made so that microwave pulses from an external source could be introduced, while the detection system was changed to permit video detection after the 30 Mc/s amplifier for the display of the recovery curve on an oscilloscope. A simplified block diagram is shown in Fig. 7.3. The required frequency stability is achieved because the basic oscillator in the equipment is quartz crystal controlled at a fundamental frequency of 30 Mc/s. The frequency of the signal klystron is locked to a harmonic of this oscillator and the signal passes to a resonant cavity from which it is reflected to the 30 Mc/s pre-amplifier. Like the signal klystron, the local oscillator signal is phase-locked to the crystal-controlled oscillator but 30 Mc/s away, so that phase-sensitive detection at this intermediate frequency can take place after the main 30 Mc/s amplifier. The magnet modulation coils are fed from the 33 c/s power oscillator. The returning resonance signal therefore has a component at this frequency, which is amplified in the receiver containing a lock-in amplifier, and the first derivative of the absorption or dispersion signal may be selected and finally drawn on the pen recorder.

The high-power microwave pulses are derived from an English Electric K350 klystron working in pulsed operation. This is a two-resonator valve capable of producing about 2 W CW and it requires a supply of 800 V at 80 mA. As used at present, the voltage between the anode and cathode of the signal klystron is set to be just below a klystron mode, and voltage pulses sufficient to bring the tube into the oscillating condition are superimposed on this. The pulse generator (Nagard Type 5002C) drives the grid of a series valve in the HT line. This simple system permits the use of microwave pulses variable in width between 10 μsec and 200 msec, having a fall time of about 1 μsec. The peak power is of the order of 2 W and repetition rates for the pulses are widely variable. This microwave power is fed into the main waveguide run through a 3 dB coupler.

Some form of suppression must be applied to the microwave signal returning to the detector so that, in the event of serious reflection from the cavity, the high pulse power is not allowed to impinge upon the mixer crystals, yet the monitor signal itself must not be appreciably attenuated. Sufficient suppression is obtained in this apparatus by using a single waveguide-mounted diode switch inserted in the run immediately preceding the mixer crystals. The diode normally operates with a positive bias and therefore in the conducting condition, causing little attenuation. Negative voltage pulses superimposed on that bias cause the switch to attenuate the microwaves by about 25 dB for the duration of the pulse. This suppression is usually sufficient to protect the mixer crystals. It is not necessary to apply further suppression to the pre-amplifier, although facilities are available in the unit for this purpose; its recovery time is 1 or 2 μsec and saturation would only affect measurements of very short relaxation times.

The pulse generator provides the synchronized positive pulses for the klystron supply and negative pulses for the suppression; in addition, it provides triggering pulses for the oscilloscope at twice the repetition frequency of the other pulses so that an unambiguous base line may be drawn along with the relaxation curve, on the oscilloscope (see p. 115).

The recovery of the absorption from saturation is monitored by extracting the signal after the 30 Mc/s amplifier, diode detecting and displaying the resulting transient on a Tektronix 535A oscilloscope. A photograph of the complete apparatus is shown in Plate 1 (facing p. 120).

Cryostat and resonant cavity assembly

Again we shall describe a system which is at present in use. Its inclusion here is simply intended to act as a guide to those who lack experience in the use of this type of equipment. The resonant cavity operates in liquid helium and must therefore be suspended some distance from the rest of the waveguide assembly, which is, of course, at room temperature. An outline of the system is shown in Fig. 7.4. The resonant cavity is cylindrical and resonates in an H_{111} mode. The base is in the form of a choked plunger, on which the sample is mounted. Fine tuning is provided by means of a quartz rod; the insertion of this rod into the cavity is controlled by means of a drive passing through the cryostat head. It is desirable that the coupling of the cavity to the waveguide shall be variable even at liquid helium temperatures, and a mechanism is employed similar to that described by Gordon.[10] A PTFE wedge is driven up and down a section of waveguide which, when empty, is beyond cut-off. This is immediately above the circular coupling iris, and the wedge is driven by means of a rod passing through the cryostat head. The wedge is tapered to reduce reflections and it alters the length of the cut-off guide presented to the microwaves and hence changes the reflection coefficient of the cavity. The necessity for variable coupling arises since, with a reflection cavity, it is important to operate close to critical coupling for maximum sensitivity (Feher[1]), yet not too close or the absorptive and dispersive components of the reflected signal may not be separable by tuning. The method adopted here provides an adequate variation in coupling for most purposes.

The waveguide, connecting the head to the cavity, and the control rods are of stainless steel to reduce heat leaks by conduction, and heat baffles are provided near the top of the cryostat. In this particular piece of equipment a double glass Dewar arrangement is used, the inner containing liquid helium and the outer liquid nitrogen. Metal Dewars have certain advantages over glass but are more costly. The temperature of the sample is measured by means of an Allen-Bradley carbon resistor, previously calibrated against

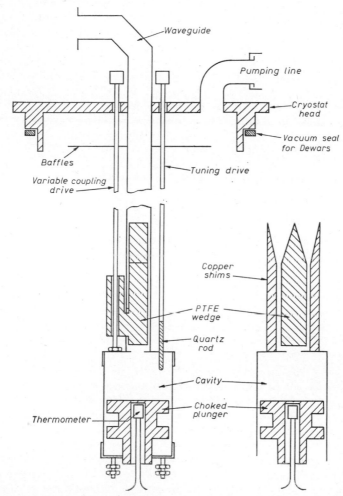

Fig. 7.4. Typical cavity and cryostat assembly showing the variable coupling mechanism and tuning rod.

a helium vapour pressure thermometer, located in a recess at the base of the cavity immediately below, and in good thermal contact with, the specimen.

If liquid helium enters the cavity, good thermal contact with the specimen is ensured. However, the dielectric constant of helium varies with temperature and hence alters the resonant frequency of the cavity, possibly outside the range of tuning available when using a quartz rod. Also, above the lambda point $(2 \cdot 2°K)$, bubbling of the liquid around the coupling iris can cause rapid fluctuations in the coupling coefficient and so introduce instabilities in the detected signal. In our laboratory it is the practice to exclude liquid helium from the cavity by using a latex sheath, long enough for the neck to be well above the liquid-helium level. These sheaths are made from pre-vulcanized latex (Dunlop compound A330), building up each by painting several layers of the liquid latex on a pyrex former of the required diameter. It is usually a fairly loose fit over the cavity and waveguide, and may be secured by an elastic band above the helium level. This sheath is thin and does not appear appreciably to lengthen the time taken for the specimen to reach thermal equilibrium at a given helium temperature. Again, this is not the only method which can be used but it is certainly a convenient one.

Temperatures between $4 \cdot 2°K$ and $1 \cdot 6°K$ can be obtained by pumping on the space above the liquid helium, using a fairly large capacity rotary pump, regulating the pressure by means of a needle valve or manostat in the pumping line.

The chief disadvantage of this apparatus is the inability to change specimens or to alter specimen orientation within the cavity once the low temperature has been achieved. Such facilities can be included at the expense of further mechanical complication.

Procedure and experimental techniques

A low-voltage klystron (not shown in Fig. 7.3) frequency modulated at 50 c/s is used in conjunction with a crystal-video detector to display a klystron mode in the initial setting up and tuning of the cavity. This procedure is desirable if large samples with a high dielectric constant are used (as, for example, in the case of ruby). At the operating temperature, any E.S.R. line of interest can be displayed on an oscilloscope, using the low-power monitor signal from the klystron and the 50 c/s field-sweep video technique. The cavity is tuned and coupled for minimum second-detector current when off magnetic resonance, and then de-coupled slightly and re-tuned until the video display of the line is observed to be absorptive by its shape. The amount of de-coupling necessary to remove all traces of dispersion depends on the intensity of the absorption, a very intense line needing much more de-coupling than a weak one.

The monitor power level is then reduced until no saturation of the resonance line occurs on collapsing the sweep to the centre of the line. The sweep is then removed, the microwave pulses applied, and the recovery of the saturated line observed on an oscilloscope using a suitable linear time base. The repetition frequency of the pulses is adjusted so that the recovery of the line is completed before the second trace is drawn on the oscilloscope, this then acting as a base line to which the recovery is tending. The recovery curve can be photographed using an oscilloscope camera, and the traces analysed at leisure. Other methods for obtaining the time constant directly are referred to below (see p. 122).

When used as a conventional E.S.R. spectrometer to display the derivatives of lines, the apparatus is capable of high sensitivity, but it is found necessary to use a sample containing about 10^{16} spins for relaxation-time measurements. A sufficiently intense signal is then obtained so that the recovery time constant can be determined within an error of about 10 per cent. This error is due partly to the noise which appears on the recovery curve with consequent uncertainty in determining the mean value of ordinates on the trace, and partly to uncertainty in estimating the true position of the base line to which the curve is tending. For more intense signals, the error may be less than 5 per cent.

A system which reduces noise on the signal will also reduce the error. This is especially important since maximum information can be extracted from the recovery curve only by analysing it in detail in the region near to complete recovery where the trace is usually obscured by the spectrometer noise. Beyond a certain limit, any simple electronic filtering of noise will tend to distort the relaxation trace. A digital memory oscilloscope may here be used to some effect. The Northern Scientific model NS544, used in the authors' laboratory, is simply a fixed programme computer used as a signal-averaging device. In it, the signal from the output of the spectrometer is sampled at 1024 points uniformly spaced in time, the dwell time at each point being variable in twelve steps from 250 msec to 62·5 μsec. The action of the memory oscilloscope is to integrate the analogue voltage amplitude at the input over one quarter of the dwell time and then to convert this integrated value to digital form and store it in the memory. This is repeated at each point until the complete signal has been converted into 1024 digital amplitude values and stored. This procedure is then repeated, and a second set of amplitude values added to those already contained in the store. Provided the triggering is so arranged as to provide a coherent signal, the random amplitude noise on the signal is reduced in comparison with the coherent signal, and, assuming a Gaussian distribution of the noise amplitude, it can be shown that after N sweeps through the signal the signal-to-noise ratio is improved by a factor $\sqrt{N}$. The limit of this improvement is set by the practical difficulties of maintaining the signal for long periods of time. An example of the improvement possible is shown in Fig. 7.5.

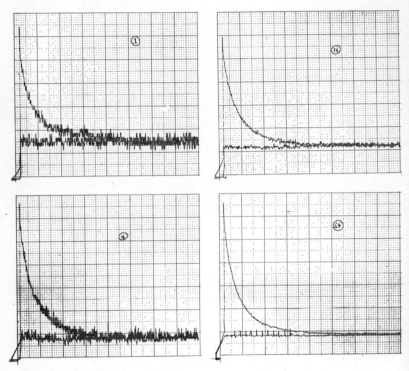

Fig. 7.5. Output traces from the memory oscilloscope after one, four, sixteen and sixty-four sampling sweeps through a relaxation recovery curve (Taylor[11]).

Alternative spectrometer arrangements

In one widely used type of spectrometer, a balanced microwave bridge is employed. Fig. 7.6 shows schematically the method employed by Standley and Wright.[4] A microwave switch is used to produce the pulses from the high-power CW klystron, and the signal is divided at a magic tee. One arm of this contains a high-quality matched load, whilst in the opposite arm there is a cavity with variable coupling similar to that described above. No signal is transmitted along the fourth arm of the tee when the cavity is critically coupled, but a signal occurs when this balance is upset by the magnetic resonance absorption in the sample. The transmitted signal is therefore a measure of the amplitude of this absorption, and can be detected by superheterodyne methods as before. The advantage of this system is that the power level incident on the detector crystal is very low, and full use can be made of the subsequent amplification available. In effect the CW power is backed off in this method and the absorption signal only is transmitted through the receiver. The same effect can also be produced using a circulator instead of a magic tee as reported, for

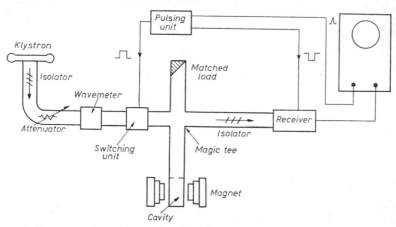

FIG. 7.6. Balanced bridge spectrometer of Standley and Wright.[4]

example, by Davis *et al.*[12]; in this case the whole of the available power is incident on the cavity instead of only one half. Pace *et al.*[13] and Bowers and Mims[6] used an E/M tuner in one arm of a magic tee in order to balance out the signal reflected from the cavity when off magnetic resonance. This method obviates the need for a variable coupling cavity, but according to Gordon[10] it is more sensitive to microphonics in the long arms of the bridge.

Scott and Jeffries[3] used a transmission cavity. This does not require a balanced bridge, and also has the advantage that the observed signal is strictly proportional to χ'' and hence to the microwave absorption in the sample. There is no possibility of difficulties arising through traces of dispersion in the signal, as can happen in a spectrometer employing a reflection cavity.

The nature of the detected signal

In an E.S.R. spectrometer one observes the variation in detected power which occurs when the applied magnetic field is varied and produces magnetic resonance in the sample. When phase-sensitive detection is used, either the absorptive or the dispersive component of the resonance signal may be selected simply by adjustment of the phase of the reference voltage. When the spectrometer is used in relaxation experiments, however, the detector is usually sensi-

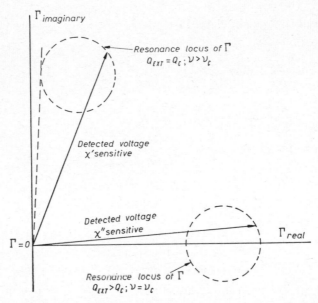

FIG. 7.7. Chart of reflection coefficient, Γ, for a variable coupling circuit. Sample resonance causes a small excursion of the reflection coefficient, shown by the circles. Slight mismatch gives sensitivity to changes in the real part of Γ, which is proportional to χ''; slight detuning gives sensitivity to the imaginary part of Γ, which is proportional to χ' (Gordon[10]).

tive only to the *amplitude* of the returning signal, and one must ensure that this signal does not contain any dispersive component. This is important since the theory at the beginning of this chapter only relates the change in absorption, and not dispersion, to the parameter T_1. When a transmission cavity is used the signal is always purely absorptive. With the more usual reflection cavity, one normally works near critical coupling, with the cavity tuned to the signal frequency; in this case the detected signal contains both absorptive and dispersive components. Gordon[10] showed that the two contributions could be separated by introducing a deliberate unbalance. The spectrometer output is made sensitive only to absorption by tuning the signal frequency to the centre of the cavity resonance (or vice versa) and then adjusting the coupling to give a small mismatch; it is made to give an output proportional to the dispersion by adjusting the coupler to exact critical coupling and then unbalancing by a small de-tuning of the signal frequency from the cavity resonance frequency. This is best illustrated by means of a diagram showing the real and imaginary components of the reflection coefficient (see Fig. 7.7). The required de-coupling can readily be achieved with the variable coupling mechanism first used by Gordon and briefly described above, but the same effect may also be produced by using a slide-screw tuner in front of the cavity. Ideally, alteration of the position of the probe along the waveguide effectively alters the reactance and hence the resonant frequency of the termination, while alteration of the depth of penetration of the probe alters the reflection coefficient and hence the degree of coupling. It should be noted that the unwanted part of the signal is never completely removed unless homodyne detection is used, but it may be reduced to a second-order effect.

Variations on the pulse-saturation technique

An early paper on the pulse-saturation technique (Giordmaine *et al.*[14]) describes an apparatus using two klystrons operating at different frequencies. These are swept synchronously through the E.S.R. absorption line, the sweeps being repeated at intervals which are long compared with T_1. If the klystron which reaches the resonance first in time is of sufficient power to saturate the line, then the second, low-power klystron, arriving at a known time later, can be used to monitor the degree of saturation remaining. By varying the frequency difference or the sweep-rates of the two klystrons, the recovery can be plotted out and a value for T_1 obtained in the usual manner.

One major disadvantage of the conventional pulse-saturation method is that the monitor power must be kept sufficiently low so as not to interfere with the recovery of the spin system, thus requiring the detection of a very weak signal. This problem becomes particularly severe when measuring long relaxation times, when the power level may have to be less than 10^{-8} W, or when using dilute specimens. The method reported by Baker and Ford[15] overcomes this difficulty. This method might be described as an extension of the continuous saturation technique to the instantane-

J

ous measurement of population during recovery. A high-power microwave field is applied to the system for a time sufficient for the spins to be saturated. It is then removed for a known time τ_1 whilst a partial recovery of the spins occurs, after which time the population difference $n(\tau_1)$ is given by $n(\tau_1) = n_0[1 - \exp(-\tau_1/T_1)]$. It is then switched on to complete the saturation, during which time the transient behaviour of the absorption signal is observed on an oscilloscope (Fig. 7.8a). The difference between the height of the observed signal when the power is first turned on again and the height after the spins have reached their maximum

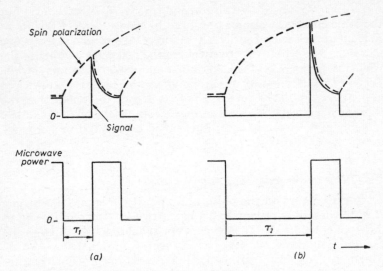

FIG. 7.8. Spin polarization, signals and microwave power as a function of time during relaxation-time measurements using the technique of Baker and Ford.[15]

saturation is proportional to $n(\tau_1)$. Next the experiment is repeated with longer times τ_2, τ_3, etc. (Fig. 7.8b), including a measurement at $\tau \gg T_1$ in order to obtain n_0. A graph of $(n_0 - n)$ against τ will then give the form of the recovery from which T_1 may be found. The advantage here is that the power level during recovery may be reduced to an arbitrarily low value, but may be as high as desired while the signals are actually being observed.

Switching may be achieved by a diode switch or, if T_1 is long enough, by removing and then returning the magnetic field to resonance. The latter method has the advantage that relaxation times may be measured at fields other than the resonance field.

Plate 1. The spectrometer described in Chapter 7. On the left can be seen the Decca X3 unit.

This technique overcomes the difficulty inherent in all resonant methods of measuring T_1, namely the inability to measure the variation of T_1 with applied magnetic field due to the limited frequency ranges of microwave equipment. It can only be used for specimens with long relaxation times as it is difficult to switch a magnetic field to and fro in less than a few seconds. Baker and Ford used this method for measurements on neodymium-doped lanthanum fluoride, and obtained a value for T_1 of 23 sec at 0·2°K. They measured T_1 in fields ranging from 100 to 2100 gauss, using a single microwave frequency of 9·7 Gc/s. At higher temperatures, T_1 was too short to switch the field, and the microwave power was switched instead using waveguide-mounted diodes.

Several workers have reported an extension of the pulse-saturation technique which enables cross relaxation and spin diffusion to be observed directly. The saturating pulse is applied at one frequency, and the absorption monitored at another. Under favourable conditions, the build-up of energy at the monitored frequency can be observed while the pulse is being applied, followed by spin-lattice relaxation after the pulse has terminated. A bimodal cavity is usually employed in order that sufficient power can be coupled into the 'pumped' transition. Bowers and Mims[6] used this method to investigate the diffusion of energy throughout the inhomogeneously-broadened line in nickel fluosilicate. The pulse was applied at the centre of the resonance line, and the monitor was set 200 Mc/s away from this. The rise in spin temperature of the monitored transition during the pulse was clearly observed.

Cross relaxation between hyperfine components has also been investigated by several authors. (This is really an Overhauser effect, since the populations of different nuclear levels are upset by saturating one E.S.R. transition.) Unruh and Culvahouse[16] observed cross relaxation between the hyperfine lines of cobalt in lanthanum zinc nitrate, while Solomon[17] observed cross saturation between resonances in the Mn^{2+} and Fe^{3+} spectra in MgO separated by as much as 50 linewidths. Steady-state cross-saturation measurements as well as the above type of transient cross-saturation measurements were reported by Solomon. Taylor et al.[18] compared measurements made on Co in LaZn double nitrate using both the conventional pulse-saturation method and also a double-frequency (Overhauser) experiment. Significantly non-exponential recoveries were often obtained for measurements made at a single frequency, whereas very clear single exponentials resulted from measurements made using a bimodal cavity. This single exponential was always in good agreement with the latter portion of the single-frequency decay; this behaviour was attributed to spectral-diffusion processes within the inhomogeneously-broadened lines.

The two-frequency type of experiment, therefore, presents a means of measuring cross-relaxation and spin-diffusion times, in certain cases, which can only be inferred from lengthy pulse-width-variation methods using the conventional single-frequency method.

Analysis of experimental data

We have so far explained how we produced the output from the pulse-saturation spectrometer displayed as a recovery curve on an oscilloscope, or perhaps drawn out on an X-Y recorder. We now require to extract as much information as possible from this curve. We assume that the receiver output is proportional to the instantaneous population difference $(n_0 - n)$ so that the amplitude of the absorption signal y at any time t after the end of the saturating pulse is given by

$$y \propto (n_0 - n) = \text{const} \exp(-t/T_1) \qquad (7.4)$$

We shall first discuss how the parameter T_1 may be abstracted from these data.

One of the simplest methods is to compare the trace directly with a standard set of exponential curves produced on a transparent film which can be placed over the face of the oscilloscope. Suitable adjustment of gain control should bring the two curves into coincidence (the absolute magnitude of the absorption being of no importance), and T_1 thereby obtained. Alternatively, electronically generated exponential curves are displayed simultaneously with the recovery curve on a double-beam oscilloscope, and the gain adjusted for coincidence. In a modification of this method (Schultz and Jeffries[19]) the generated exponential voltage is applied to the X plates of the oscilloscope and the time constant of the generator is adjusted until a straight line is displayed on the screen. The relative amplitudes will then determine only the angle of the line on the screen and the only variable requiring adjustment is the time constant, as opposed to the previous method where two independent variables are involved.

These methods work perfectly well so long as only one time constant is involved or when two very different time constants appear in one recovery. The presence of more than one such constant may not show up if the deviation from a pure exponential is small. The following method of analysis is much more tedious, but it is always to be preferred unless one is quite sure that only one time constant is present. It enables a deviation from the condition of a single exponential to be readily and certainly detected, and in favourable circumstances may enable more than one time constant to be determined from a single recovery.

From equation (7.4) it follows that

$$\ln y = -t/T_1 + \text{constant} \qquad (7.5)$$

and a graph of $\ln y$ against t should yield a straight line from the slope of which T_1 is determined. In the usual method, one picks off pairs of points from a photograph of the oscilloscope trace (for example by projecting the negative on graph paper and drawing in the curves) and plots these semi-logarithmically, as shown in Fig. 7.9. The figure shows a condition in which the relaxation parameter T_1 is determined unambiguously.

Simple exponential decays resulting in linear semi-log plots mean that the spin system has only one decay constant. This is certainly not always true and when more than one time constant is present this is manifested in semi-log plots which may have the form shown in Fig. 3.4; a pronounced curvature of the plot will be found in those cases where the time constants present are of the same order of magnitude. In such cases it is only possible to make order of magnitude estimates of the time constants from the asymptotic gradients of the semi-log plot of the recovery. When the time constants are very different from one another, the curve is simpler to analyse since each time constant will predominate over the other at different time periods. It may therefore be possible to separate the two contributions by choosing an appropriate oscillograph time-base speed. The semi-log plot will then be separable into two fairly linear portions. Any slight error in estimating the base line to which the decay is recovering will show up as a 'lift' of the initial part of the plot, and this can lead to doubt as to whether the recovery curve has a long tail on it or not. This error is likely to occur when one is measuring very long relaxation times of the order of hundreds of milliseconds, since any slight change in d.c. level between successive sweeps of the oscilloscope will shift the apparent base line to which the recovery is tending. The importance of a well-stabilized d.c. level is therefore apparent.

The problem of the interpretation of recovery curves due to more than one time constant has been discussed by Standley and Vaughan[20] and illustrated by the work carried out on the relaxation of the Cr^{3+} ion in Al_2O_3 (ruby). A theoretical study of this problem was carried out by Donoho,[21] It was shown that relaxa-

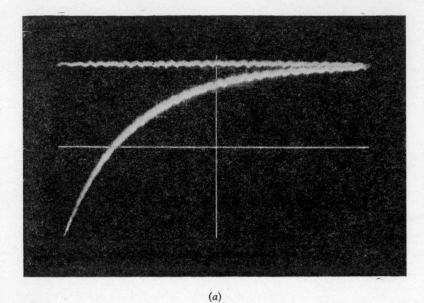

(a)

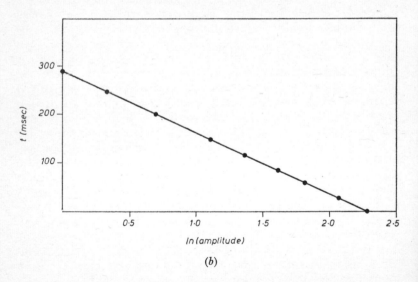

(b)

Fig. 7.9. The experimental recovery curve shown in (a) is plotted semi-logarithmically in (b). This is seen to be described closely by a single time constant.

tion in this four-level system involved three independent relaxation parameters thus

$$\frac{n}{n_0} = 1 + A_1 \exp(-t/\tau_1) + A_2 \exp(-t/\tau_2) + A_3 \exp(-t/\tau_3) \quad (7.6)$$

The parameters $\tau_{1,2,3}$ are time constants, not always simply related to the $T_1^{(ij)}$ which may be defined, for a pair of levels i and j, by the equations of Chapter 1, and $A_{1,2,3}$ are the relative amplitudes of the contributions. The numerical calculations of Donoho show that A_3 is generally negligibly small compared with A_1 and A_2 and it is omitted from the following discussion.

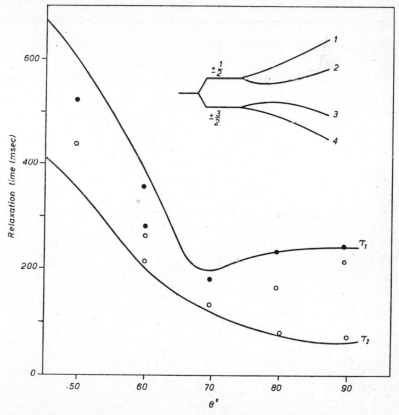

FIG. 7.10. Comparison of theoretical and calculated relaxation data for the $(3 \rightarrow 4)$ transition (indicated in inset) for ruby as a function of crystal orientation (Standley and Vaughan[20]). —— are the Donoho[21] curves and ● the values calculated from them. ○ are the experimental points. Only at 60° does the analysis predict that two time constants should be resolved. At other angles a single intermediate value is expected.

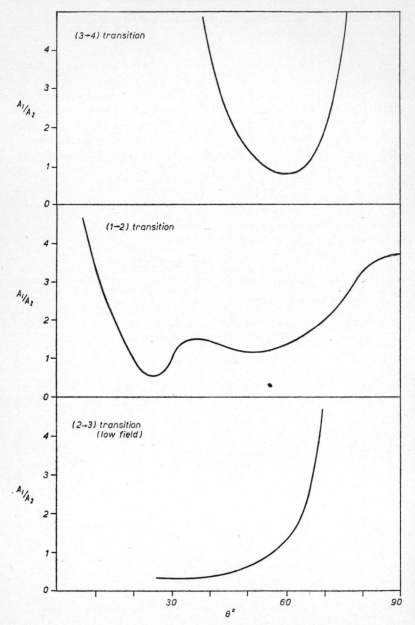

FIG. 7.11. Relative amplitudes of the contributions from τ_1 and τ_2 to the relaxation processes, calculated from the data given by Donoho[21] for various transitions in ruby.

In order to discover the experimental implications of Donoho's analysis, Standley and Vaughan computed the form of the recovery curves to be expected, using Donoho's numerical data. The time constants which one would determine experimentally were obtained from semi-log plots of these computed recovery curves, in the manner described above. The results are shown in Fig. 7.10 while Fig. 7.11 shows the A_1/A_2 ratio calculated from Donoho's data. These figures refer to the 3–4 transition, indicated in Fig. 7.11. The important result is that only in certain cases does one find that the longest time constant determined in this way is equal to τ_1. When $A_1 \gg A_2$, the observed recovery should follow an exponential law and the semi-log plot should give τ_1 unambiguously; a component due to τ_2 may well not be distinguished. The case $A_2 \gg A_1$ is the reverse of this. The important condition is when A_1 and A_2 are similar in magnitude. If τ_1 is also similar to τ_2, the curve is apparently near exponential and characterized by a parameter intermediate between τ_1 and τ_2. If τ_1 and τ_2 are very different, the two time constants may be resolved. In Fig. 7.10 some of these expected experimental points are shown together with those actually determined using the apparatus and procedure described in this chapter.

As explained in Chapter 3, the presence of cross-relaxation processes may also result in recovery curves which are governed by two time constants, the relative amounts of which depend on the duration of the saturating pulse. The two components may usually be separated by varying the pulse width. If a sufficiently long pulse is used, then all cross relaxation takes place during the pulse and the monitored recovery has a good chance of exhibiting only a single time constant, which may often be taken as T_1.

An allied problem was also the subject of experimental and theoretical study by Standley and Tooke.[22] The material used in this investigation was chrome alum diluted into the isomorphous aluminium alum and containing between 0·1 and 6·0 atomic per cent Cr^{3+}. At room temperature the E.S.R. spectrum is a simple three-line one, which would be expected for an ion with $S = \frac{3}{2}$ and with a moderate zero-field splitting characterized by the parameter D in the spin-Hamiltonian. A phase change occurs at about $160°K$, below which temperature at least three different D values appear to exist within the crystal. Moreover, the value of D is dependent on the degree of magnetic dilution. Thus under the combined influence of dilution and low temperature, the values of D in the crystal are very small but remain finite. This causes the spectral lines to cluster and eventually the individual components are not resolved and a single rather broad resonance line is observed. Thus, in this material, the E.S.R. line is inhomogeneously broadened although the source of the broadening is different from that usually found.

On saturating the centre of this composite line, general cross-relaxation processes (see Chapter 3, p. 53) share the energy through the entire line; this becomes completely saturated if adequate incident power is available. The experimentally determined relaxation parameter must therefore be interpreted as a time constant which is characteristic of the recovery of the spin system as a whole. Standley and Tooke attempted to consider the

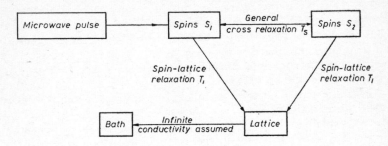

FIG. 7.12. A simple theoretical model indicating the cross relaxation between spins S_1 and S_2 and their relaxation to the lattice and to the bath (Standley and Tooke[22]).

effect of this cross relaxation through the line with the following simplified model. The spin system was considered to be divided into two groups, spins S_1 which were saturated by the microwave pulse and the remainder S_2 which became excited only through general cross relaxation. The simple physical model is then as shown in Fig. 7.12. After saturation, the excited spins S_1 can lose energy and decay to their ground state through two distinct paths, one direct to the lattice and the other an indirect two-stage process via the spins S_2. The former process involves the normal spin-lattice relaxation and is characterized by the usual time constant T_1. The indirect process involves both cross relaxation between the two spin groups and spin-lattice relaxation of the spins S_2. These two processes are assumed to have characteristic times T_s and T_1 respectively; for complete generality the spin-lattice time constant should not be considered the same for both S_1 and S_2, but a single T_1 is not an unreasonable first-order approximation.

The rate equations are then very similar to those discussed in Chapter 3, pp. 45—49, where T_{21} corresponds to T_s. The recovery is described by the equation

$$n_0 - n = n_0 \exp(-t/T_1) \left[L + M \exp(-2t/T_s)\right] \qquad (7.7)$$

where L and M are constants for a particular set of experimental conditions, including the saturating pulse width Δ. The deviation from a theoretical single exponential form is readily apparent as is the influence of the parameter Δ on the recovery. The experimental recovery curve is in effect a graph of n as a function of time. In making a semi-log plot, values of $(n_0 - n)$ are taken from the experimental curve and $\ln(n_0 - n)$ is plotted against t. For the ideal single exponential recovery we have equation (7.5)

$$\ln(n_0 - n) = -\frac{t}{T_1}$$

Thus the gradient is $-1/T_1$ and the relaxation time is determined. By a similar argument from equation (7.7) we have

$$\frac{d}{dt} \ln(n_0 - n) = -\frac{1}{T_1} - \frac{(2M/T_s)\exp(-2t/T_s)}{L + M \exp(-2t/T_s)} = \frac{1}{T_{1e}} \qquad (7.8)$$

where T_{1e} is the time constant obtained from the gradient of the semi-log plot at any time t after the termination of the saturating pulse.

Equation (7.8) may be written simply as

$$\frac{1}{T_{1e}} = \frac{1}{T_1} + X \qquad (7.9)$$

where X is an experimental parameter which depends on T_1, T_s and Δ. The magnitude of X thus indicates the influence of the cross relaxation upon the measured time constant. Standley and Tooke determined X for a series of values of Δ, using a value of $T_s = 1$ msec, in accordance with other experimental data. The results are shown in Fig. 7.13 together with the measured value of $T_1{}^{-1}$ at $4 \cdot 2^\circ$K, for comparison. During the period immediately following the saturating pulse, the cross relaxation plays a dominant role, even when pulses as long as 2 msec are used. When pulse widths of the order of 100 μsec are employed (and such pulses have not been uncommon in the past) the value of X diminishes to 10 per cent of $T_1{}^{-1}$ only after some $1\frac{1}{2}$ msec. These results highlight the necessity for the careful study of recovery curves close to the time when recovery is complete. One is then working against the noise level of the equipment and the use of a time-averaging system such as that already described on p. 115 can increase, by a factor of 10, the time available for measurement by increasing the effective signal-to-noise ratio.

It is thus important to decide just how unambiguous is the information obtained. In other words, one must always pose the question, do we obtain a true measure of the spin-lattice relaxation time T_1 from such experiments, or is there room for doubt as to the correctness of the interpretation? Mims, Nassau and McGee[23] have drawn attention to another source of error which may occur

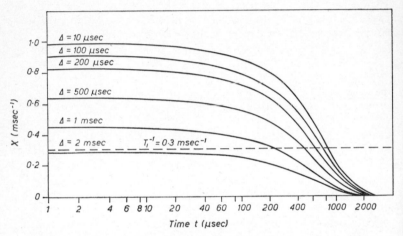

FIG. 7.13. Effect of pulse width Δ on the observed recovery curve. X is the difference between the instantaneous and true spin-lattice relaxation rates. The time t is measured from the end of the saturating pulse.

when pulsed microwave techniques are used to study the relaxation in inhomogeneously-broadened lines. In such a case the pump pulse may succeed only in burning a hole in the line and the monitor signal will show the filling in of this hole from the sides. The recovery trace is then strongly affected by this diffusion of energy and the trace form depends markedly on the diffusion function (see p. 151) and on the shape of the hole originally burned in the line. From their analysis of the spectral diffusion problem Mims *et al.* arrived at the conclusion that the relaxation trace should have the hyperbolic form:

$$\frac{y}{y_0} = \frac{1}{1 + at}$$

Such a form of the trace was found by Mims *et al.* in several cases during their measurements on cerium and erbium ions in calcium tungstate. Clearly their experience raises the question: Which part, if any, of a relaxation trace which at first sight appears to consist of a fast followed by a slow period can legitimately be ascribed to true spin-lattice relaxation? One simple test is to vary the intensity and the duration of the saturating pulse and to observe if there are any changes in the recovery curve, particularly during the slow period. If none is apparent, this is indirect evidence that the spin-lattice relaxation is controlling the decay rate at the beginning of the slow period and that further diffusion within the line is having a second-order effect. The measured parameter may still not be exactly the true spin-lattice relaxation time but it will be close to it. If the slow period parameter is found to vary with the saturating pulse conditions and not to reach a steady value, then it is most unwise to regard these measurements as giving spin-lattice relaxation times. In the work of Mims *et al.* this was found to be the case for the cerium specimens and no lattice relaxation time was measured for these. We must again emphasize that many published data on spin-lattice relaxation times do not contain any statement or evidence that the effect of pulse widths has been studied, and indeed many early experiments employed pulses of only a few microseconds duration.

Advantages of the pulse-saturation technique

At the time of writing, this is the most widely used of the several methods which are available. It is extremely flexible in its application but requires only fairly simple modifications and additions to a standard spectrometer. It does not require modulation or pulse operation of the magnetic field. A single microwave pulse is used, pulse widths varying from a few microseconds to about $\frac{1}{10}$th second and these are easy to produce; one contrasts this with the more difficult sequence of micro- or nano-second pulses required in, say, the spin-echo method (§8.1).

The simple relaxation behaviour characterized by a single relaxation parameter is only rarely observed, and there will normally be present such complicating phenomena as cross relaxation, spin diffusion or phonon bottlenecks for example. However, in many cases it is possible to identify these complicating processes, albeit only indirectly, using this technique and in favourable circumstances it is then possible to obtain T_1 provided that sufficient parameters can be experimentally varied. Both the non-resonant method and the continuous saturation method measure the net relaxation which is taking place by any possible route and so may well provide less information in involved cases than does the pulse-saturation method.

However, the difficulties and doubts which the above arguments raise must be borne in mind when one is interpreting published data, as we have attempted to do in Chapter 9 and Appendix 2.

§7.2

USE OF D.C. MAGNETIZATION MONITOR

The method to be described in this section was used successfully by Damon[24] in 1952 and by Bloembergen and Wang[25] in 1954 when studying relaxation in ferromagnetic metals; it was not till 1962 that Bloembergen and Feng[26] first reported the successful application of the technique to relaxation in paramagnetic salts. Their operating frequencies were in the 9 and 24 Gc/s bands and the method was subsequently used at 35 Gc/s by Lees, Moore and Standley.[27]

The essential physics of the method is as follows. Consider a system which consists of only two spin levels associated with the magnetic quantum number $s_z = \pm \frac{1}{2}$. In an applied magnetic field H there will be a Zeeman splitting of the levels which can be made equal to an incoming microwave quantum $h\nu$. We suppose that only the two spin levels are populated significantly at the temperature of the experiment. In the absence of any microwave power, the macroscopic magnetization of the sample in the direction of H results from the greater number of spins in the lower ($-\frac{1}{2}$) level. In this simple system, the ratio of the two populations is given by the Boltzmann equation

$$\frac{N_{\frac{1}{2}}}{N_{-\frac{1}{2}}} = \exp(-h\nu/kT)$$

[see equation (1.8), p. 10; N_i and n_i refer to equilibrium and non-equilibrium populations, respectively, of the level i]. When the microwave power is switched on, absorption of energy by the system occurs, and, if sufficient power is present in the microwave field, the two levels become equally populated, i.e. $n_{\frac{1}{2}}^{sat} = n_{-\frac{1}{2}}^{sat}$. The magnetization in the direction of H has thus fallen to zero. If the saturating microwave power is now removed, the original population distribution in the system—and hence the original magnetization—is regained in a finite time.

Clearly we have chosen as an example a particularly simple system which will rarely be met in practice. But it can readily be seen that a similar argument can be applied to, say, a four-level system as shown in Fig. 7.14. We write the levels as pure spin states where S_z takes the values $\frac{3}{2}, \frac{1}{2}, -\frac{1}{2}, -\frac{3}{2}$, while the populations N_i and n_i refer to unit volume of the sample. If we assume that a constant g-value exists, the magnetization, the magnetic moment per unit volume, in the direction of the applied field is

$$M_0 = [(-\tfrac{3}{2})N_1 + (-\tfrac{1}{2})N_2 + (+\tfrac{1}{2})N_3 + (+\tfrac{3}{2})N_4]g\beta \quad (7.10)$$

where β is the Bohr magneton, and the populations of the levels have their thermal equilibrium values.

Suppose a microwave signal now saturates the $(-\tfrac{3}{2}) - (-\tfrac{1}{2})$ transition between levels 1 and 2 so that now

$$n_1{}^{\text{sat}} = n_2{}^{\text{sat}} = \frac{N_1 + N_2}{2}$$

In bringing this about $(N_1 - N_2)/2$ spins are promoted from level 1

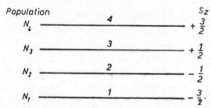

FIG. 7.14. Energy-level scheme for $S = \frac{3}{2}$. The populations N_i refer to thermal equilibrium values; in the text n_i are non-equilibrium values.

to level 2 and this results in a decrease in magnetization of the sample of

$$\frac{N_1 - N_2}{2} g\beta$$

We shall again assume a simple case where there is no cross relaxation between the four levels and the separations of levels 2, 3 and 4 are such that no absorption of the microwave power occurs except between levels 1 and 2. When the saturating radiation is removed, the original population distribution is recovered, that is, the magnetization in the direction of the applied field returns to its original value. Thus, when more than two levels in a system are populated, saturation of two of these levels does not reduce the magnetization to zero, but a *change* in magnetization still occurs and the recovery may be studied.

We may write the magnetization M_v in the sample under the influence of the saturating microwave radiation as

$$M_v = [-(N_1+N_2)+(+\tfrac{1}{2})N_3+(+\tfrac{3}{2})N_4]g\beta \qquad (7.11)$$

and, when this radiation is removed and the recovery of the thermal equilibrium populations is in progress,

$$M = [(-\tfrac{3}{2})n_1+(-\tfrac{1}{2})n_2+(+\tfrac{1}{2})N_3+(+\tfrac{3}{2})N_4]g\beta \qquad (7.12)$$

If this recovery is characterized by a single relaxation time, T_1, equation (1.21) applies to this system and we have

$$n_1-n_2 = (N_1-N_2)\,[1-\exp(-t/T_1)] \qquad (7.13)$$

Remembering that $n_1+n_2 = N_1+N_2$, manipulation of the equations (7.10), (7.11), (7.12) and (7.13) yields

$$M = M_0-(M_0-M_v)\exp(-t/T_1) \qquad (7.14)$$

Thus the change in magnetization follows the same simple exponential law, and the parameter T_1 can clearly be determined from its study. Although this was shown for a simple case, equation (7.14) is generally applicable so long as (7.13) is also applicable.

The experiment then is as follows. The specimen is placed inside a waveguide resonant cavity, or similar structure, wherein it can be subjected to a saturating pulse of microwave radiation and and can also experience the steady field H. Outside this structure is a coil system in which the varying magnetization of the specimen can set up an e.m.f., according to the normal laws of electromagnetic induction. Then, if one studies this e.m.f. as a function of time during and after the saturating pulse, one is observing a parameter which is proportional to the magnetization of the specimen, which is in its turn dependent on T_1, as we have shown. From a quantitative analysis of the recovery curves, the relaxation parameters may therefore be determined.

This experimental method differs from those of §7.1 in that no microwave signal is used in the detection of the relaxation. A relatively simple pulsed microwave saturating power supply can be employed while the detection problem is just that of measuring the small uni-directional e.m.f. generated in the pick-up coil by the changing magnetization of the specimen.

The experimental technique

A block diagram of the apparatus is given in Fig. 7.15. A pulse of microwave radiation is generated and fed to the specimen which is inside a microwave circuit and located between the poles of an electromagnet supplying the Zeeman field H. Detection takes place in the coil-amplifier-display system which is outside the microwave circuit. In their work Bloembergen and Feng[26] used as klystron sources a Varian type V58, giving $\frac{1}{2}$ W output at 10 Gc/s, and a Raytheon 2K33, giving about 50 mW at 25 Gc/s. Lees *et al.*[27] employed an Elliott 8TFK2 klystron at 35 Gc/s which was capable of supplying approximately 20 W.

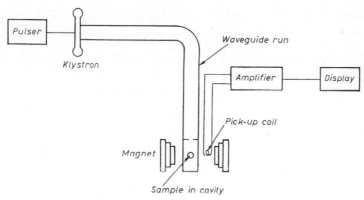

FIG. 7.15. Block diagram of the apparatus used for measuring the change in d.c. magnetization brought about by changes in energy-level populations.

The induced e.m.f. in the coil

It is possible to determine an order of magnitude of the e.m.f. which may be expected in the pick-up coil.

Consider the simplified geometry of Fig. 7.16 where there is a point dipole of moment m at the centre of the coil system. The fields H_r and H_θ are given by

$$H_r = \frac{2m}{r^3} \cos \theta, \quad H_\theta = \frac{m}{r^3} \sin \theta$$

using the c.g.s. system of magnetic units. The flux linking the coil of $\mathcal{N}$ turns is simply the flux through one hemisphere multiplied by $\mathcal{N}$. The contributions from H_θ cancel out from symmetry considerations.

K

Thus
$$\phi = \mathscr{N} \int\limits_{\substack{\text{Area of} \\ \text{hemisphere}}} H_r d\,(\text{area})$$

$$= \mathscr{N} \int\limits_{-\pi/2}^{+\pi/2} \frac{2m \cos \theta}{r^3}\,.\,2\pi r \sin \theta\, r\, d\theta$$

$$= \frac{2\pi \mathscr{N} m}{r}$$

Hence
$$\frac{d\phi}{dt} = \frac{2\pi \mathscr{N}}{r} \left(\frac{dm}{dt}\right)$$

Fig. 7.17 represents the input circuit of the coil and amplifier. If we make the resistance of the coil small compared with the input resistance of the amplifier, R represents effectively the latter quantity.

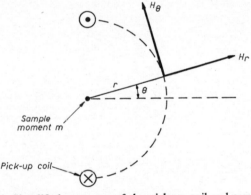

Fig. 7.16. Simplified geometry of the pick-up coil and sample. The latter is considered as a small dipole of moment m.

Then the induced e.m.f. $= L\dfrac{di}{dt} + Ri$

$$= \frac{d\phi}{dt} \times 10^{-8} \quad \begin{array}{l}\text{(employing practical units for the} \\ \text{electrical quantities)}\end{array}$$

$$= \frac{2\pi \mathscr{N}}{r} \frac{(M_0 - M_v)}{T_1} V \exp(-t/T_1) \qquad (7.15)$$

where V is the volume of the sample, assumed to be uniformly magnetized, and we have used equation (7.14) for the magnetization.

If we make the initial conditions $i = 0$ when $t = 0$, the above equation can be solved to yield

$$i = \frac{2\pi \mathcal{N}}{r} \times 10^{-8} \frac{(M_v - M_0)}{L - RT_1} V \left[\exp(-t/T_1) - \exp(-Rt/L)\right] \text{ amps}$$

(7.16)

and hence the input voltage to the amplifier is given by

$$e = \frac{2\pi \mathcal{N}}{r} \times 10^{-8} \frac{(M_v - M_0)}{(L/R) - T_1} V \left[\exp(-t/T_1) - \exp(-Rt/L)\right] \text{ volts}$$

(7.17)

Since we wish to analyse and interpret the recovery curves (i.e. plots of e as functions of t) in as simple a manner as possible, it is advantageous to choose experimental arrangements which simplify equation (7.17).

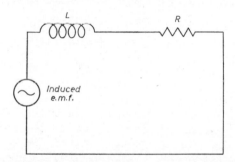

FIG. 7.17. The effective input circuit of the amplifier and coil. L represents the inductance of the pick-up coil and R includes both the coil resistance and the input resistance of the amplifier.

Firstly, the denominator contains the term $(L/R) - T_1$. In the Bloembergen and Feng experiment L was about 1 mH and R was 200 kΩ, giving L/R roughly 5×10^{-9} sec. Thus so long as values of T_1 longer than, say, 10 μsec are involved L/R may be ignored in comparison with T_1. The form of the variation of the voltage with time is conditioned by the exponential terms involving T_1 and the apparatus time constant L/R; the latter may be considered as a distorting term which may lead to incorrect values of T_1. The magnitudes just quoted show again that the distortion will be negligible for values of T_1 greater than a few microseconds, so long as an analysis of the recovery curve does not depend too strongly on that part of the curve very close to $t = 0$.

In the apparatus used by Lees *et al.*, the coil was necessarily smaller in size so that L/R was of the order 10^{-6} seconds. Since T_1 was approximately 10^{-3} seconds in that experiment, it was still permissible to ignore terms in L/R in comparison with T_1.

With these simplifications equation (7.17) now becomes

$$ e = \frac{2\pi \mathcal{N}}{r} \times 10^{-8} \frac{(M_0 - M_v)}{T_1} V \exp(-t/T_1) \text{ volts} \qquad (7.18) $$

The voltage will be a maximum, e_{max}, when $t = 0$. For a paramagnetic salt at room temperature the magnetic susceptibility is of the order 10^{-4} e.m.u. cm^{-3} and, in a field of 1000 gauss with a sample volume of 0·5 cm^3, $M_0 V = 0·05$ erg per gauss. The samples used by Bloembergen and Feng were rubies with Cr^{3+} concentration of the order 0·5 per cent, so that for these $M_0 V$ was about $2·5 \times 10^{-4}$ erg per gauss. At helium temperatures this rises to $2·5 \times 10^{-2}$ erg per gauss approximately if one assumes a Curie law for the variation of magnetic susceptibility with temperature. The coil had 8000 turns and a radius close to 1 cm and hence, assuming a value of T_1 of the order 5 msec at helium temperatures, e_{max} was calculated to be of the order 2·5 mV. Actual measurements gave a maximum induced voltage of 2 mV. The coil used by Lees had 10 000 turns and a mean effective radius of about $\frac{1}{2}$ cm, while the applied magnetic field H was of the order 10 000 gauss. The ruby samples had a concentration of about 0·1 per cent Cr^{3+}. Thus the expected e_{max} was of the order of 10 mV, but the optimum geometry of Fig. 7.16 could not be achieved and the actual induced voltage was not more than 1 mV.

The coil system

It is necessary to determine the optimum geometry of the pick-up coils carefully in order that the change in magnetic moment may induce as large an e.m.f. as possible. Since the sample becomes magnetized in the field direction, the coils must have their axes parallel to the field and as the induced e.m.f. is proportional to the rate of change of the flux, the largest number of turns gives the greatest flux linkage. However, it is quite possible to make the coil radius too large, or to place the coil too far from the sample, so that very little flux change links the more distant turns of the windings. Thus the coils should be wound close to the sample with as many fine turns as possible up to a radius where the flux linkage becomes inefficient.

In practice the radius may be limited by the requirement that the coils must be used inside Dewar flasks which have to fit within a magnet gap. Fig. 7.18 refers to the system used by Lees *et al.* where two coils were placed one on either side of the waveguide opposite to the sample. Each coil consisted of 10 000 turns of 48 s.w.g. enamelled copper wire set with 'Araldite' glue and measuring ½ inch in diameter by ¼ inch long, with a room-temperature

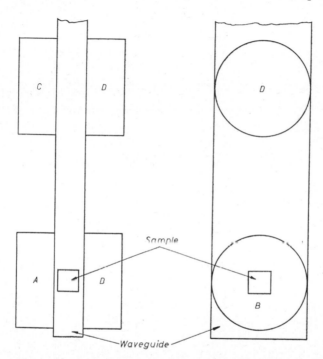

FIG. 7.18. Diagram showing the coil system used by Lees.[7, 27] The four identical coils A, B, C, D, each had 10 000 turns. A was connected in series aiding with B, and C with D, and the two outputs were opposed.

resistance of 3·5 kΩ, and mounted on a perspex support which was then located on the guide by a central stainless steel lug. The two coils were connected in series in the same sense so that the induced e.m.f.s were additive.

Since the design of the coils naturally causes them to present a large 'area-turns' to the field of the electromagnet, any variation with time in this static field causes an induced e.m.f. across the coils and hence interferes with the required signal. In the experiment one will need to observe induced voltages at least 1/10th of e_{max} and thus ideally one requires random e.m.f.s less than say

10 per cent of this. They should not therefore exceed 1 or 2 μV. For a coil of 10 000 turns, 1 cm^2 in area, this means

$$10^4 \frac{dH}{dt} \times 10^{-8} < 10^{-6}$$

or
$$\frac{dH}{dt} < 10^{-2} \text{ gauss/sec}$$

i.e. field changes at a rate of more than 0·01 gauss per second in 10 000 gauss are o be excluded. Clearly, this places an unreasonable limitation on the requirements for short-term field stability, and a compensating coil system must be used. This comprises identical coils also connected in series and mounted in a similar static field but sufficiently far away from the sample as to be outside its magnetic influence. If these coils are connected in series opposition with the first pair, the e.m.f.s induced by variations in H are to some extent balanced out, while those induced by the sample still appear across the output of the four coils. Lees found the balancing of the pairs of coils to be very critical and careful adjustment was necessary. He claimed, however, that the voltage induced by a 50 c/s modulation of the field was reduced by a factor of more than 100 through the use of the compensating coil system.

In both experiments amplification was followed by integration before display. The latter had two effects: (*a*) it removed much of the high-frequency noise so that the signal-to-noise ratio was improved, and (*b*) the integrated signal was proportional to the magnetization of the sample at any time instead of the time derivative of this quantity.

Experimental results

Published results are available only in two cases where this technique has been applied to the study of relaxation in ruby. It is clear that reliable results are obtainable. Three of the samples used by Bloembergen and Feng were later also examined at about 9 Gc/s by the pulse saturation and recovery technique described in the previous section (Standley and Vaughan[20]). Lees[7] also used both techniques at 35 Gc/s. In each case the results obtained by the two different methods agreed within their combined experimental error. Examples of the displayed signals obtained by these workers are given in Figs. 7.19 and 7.20. The latter is most interesting, for the integrated signal clearly shows an initial and abrupt rise in the magnetization when the saturating pulse is first switched on, followed by a fall and reversal of sign and subsequent recovery after completion of the pulse. The explanation of this phenomenon

$\dfrac{dM_z}{dt}$ ↑

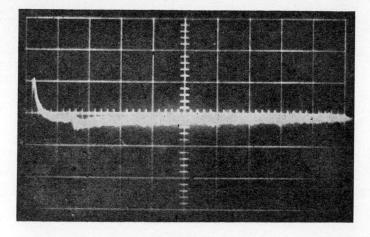

$t\rightarrow$

20 msec/cm

(a) Amplified output of pickup coils

M_z ↑

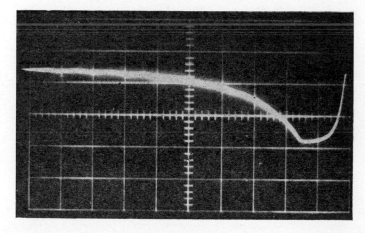

$t\rightarrow$

20 msec/cm

(b) The integrated signal

FIG. 7.19. Example of the experimentally determined magnetization curves obtained by Lees.[27] M_z is given by equation (7.14). The ruby sample contained 0·2 per cent Cr^{3+}, $T = 4\cdot2°K$, $\nu = 34\cdot2$ Gc/s, $\theta = 0°$.

is readily seen by reference to Fig. 7.21 where the spin states indicated are the dominant ones, although these are no longer pure. There is a strong cross-relaxation possibility here since $2\nu_{23} = 3\nu_{34}$. Initially there is a rise in magnetization, due to saturation of levels 3 and 4, and hence a decrease in the number of spins with $S_z = -\frac{3}{2}$. Immediately, however, and while the pulse is still on, cross relaxation occurs; for every three spins which flip down again

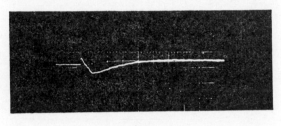

$$t \rightarrow$$
100 msec/cm

Fig. 7.20. Integrated relaxation signal (proportional to M) measured by Feng.[28] The ruby sample contained 0·04 per cent Cr^{3+}, $T = 4\cdot2°K$, $\nu = 8\cdot4$ Gc/s, $\theta = 21°$.

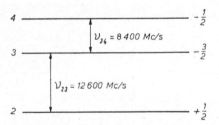

Fig. 7.21. Energy-level diagram for ruby corresponding to the case of Fig. 7.20. The probability of cross relaxation is high since $3\nu_{34} = 2\nu_{23}$. (Level 1 has been omitted for clarity.)

to level 3, two are promoted from level 2 where the dominant spin state is $|+\frac{1}{2}\rangle$. If the saturation radiation is maintained for long enough, the populations of the three levels become constant and consideration of the rate equations (pp. 133−134) shows that the changes in population can be sufficient to change the sign of the magnetization. The observed signal in Fig. 7.20 after the pulse is due to the common recovery of the population of the three levels 2, 3 and 4 coupled by cross relaxation.

Figs. 7.19 and 7.20 demonstrate one advantage of this technique. Since no microwave signal is used in the detection, it is possible to observe the change in the populations *during* the saturating pulse, a feature not found in most other methods.

§7·3

USE OF FARADAY ROTATION AS THE MONITOR

When a beam of plane polarized light passes through certain materials in the presence of a magnetic field, the plane of polarization is rotated. This is known as the magneto-optical Faraday effect and in paramagnetic crystals it arises from differences in the populations of various spin states. In the presence of a suitable externally applied magnetic field, microwave radiation upsets the equilibrium population of these spin states and consequently electron spin resonance may in principle be detected by its effects on the Faraday rotation. Such an effect was first predicted by Kastler,[29] and a quantum-mechanical theory subsequently developed by Opechowski.[30] Furthermore, since it is the *instantaneous* difference in population between the spin states which determines the instantaneous rotation, the *change* in rotation with time, after a resonance has been saturated, will give a direct measure of the rate at which the populations of the spin states return to their equilibrium values. With suitable detection and presentation equipment, the Faraday effect can thus be used to study spin-lattice relaxation.

This technique was used by Rieckhoff and Griffiths[31] to measure T_1 in single crystals of neodymium ethylsulphate. The Faraday rotation is very large for this material ($0·6°$ per mm pathlength per gauss at $2·1°$K for mercury green light, $\lambda = 5461$ Å). The measurements were carried out at temperatures between $1·4°$K and $4·2°$K, in magnetic fields in the range 800 to 2600 gauss, using pulses of X-band microwave radiation to disturb the equilibrium of the spin system. The experimental arrangement used by these authors is shown in Fig. 7.22. The microwave circuit can be rudimentary, comprising only a klystron source of microwave pulses which are applied to the sample in a resonant cavity. It was found convenient by the above authors to include a crystal-video display of the microwave resonance signal in order to adjust the magnetic field to its resonant value. The green light from a mercury arc was polarized and focused through an axial hole in the magnet onto the crystal which was mounted in the resonant cavity, immersed in a helium bath. The light emerging from the crystal then passed through an analyser, and fell on the photocathode of a photomultiplier, the output from which was displayed on an oscillo-scope. The change in intensity of light falling on the detector as the magnetic field was varied indicated the presence of an E.S.R transition. Observation of the change in intensity as a function of time after a saturating microwave pulse had been applied to the sample on resonance enabled T_1 to be evaluated in a manner analogous to that described in §7.1.

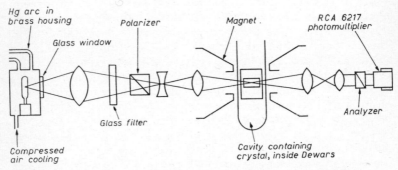

Fig. 7.22. Experimental arrangement of Rieckhoff and Griffiths[31] for measuring the Faraday rotation produced by a paramagnetic crystal.

The same authors also carried out continuous saturation meas-urements using this technique. A saturation parameter S' [see equation (6.7)] can be defined as

$$S' = \frac{\rho_0 - \rho}{\rho_0}$$

where ρ_0 is the Faraday rotation in the absence of microwave radiation, and ρ the rotation in the presence of such a field. The variation of S' with microwave power input, P_i, to the cavity was investigated, and an empirical relationship between them found of the form

$$S'(P_i) = \alpha P_i^n / (1 + \alpha P_i^n)$$

where $n \sim \frac{1}{2}$ for $S' > 0.1$ and $n \sim 1$ for $S' < 0.1$. No interpreta-tion of the form of this relationship was attempted.

An incidental but nevertheless important observation was made in one particular experimental run. The optical quality of the crystal image was sufficiently high to allow visual measurements of a semi-quantitative nature on the saturation factor S' as a function of position in the crystal for a given microwave power input to the cavity. In the absence of microwaves the crystal appeared uniformly dark with the analyser set for minimum intensity. With microwave power incident, however, the upper and lower regions of the circular crystal image darkened for distinctly different settings of the analyser, indicating different degrees of saturation in different regions of the crystal. At about $1 \cdot 73°$K, the difference between these two extreme degrees of saturation was about 18 per cent ($S' = 55 \cdot 3$ per cent at the top and closer to the helium bath while $S' = 73 \cdot 5$ per cent at the bottom). The spatial separation of these two regions was approximately $0 \cdot 15$ cm. An order-of-magnitude calculation rules out the possibility that this spatial gradient in S' is due to nonuniformity of the radio-frequency field in the cavity, and it is concluded that this effect is due to the heating of the lattice by the relaxing spin system. Under steady-state conditions, the thermal conductivity of the crystal lattice limits the rate at which energy can be transferred from the spin system to the helium bath. The spatial gradient in the 'spin temperature' is thus a consequence of a physical temperature gradient in the lattice.

Daniels and Rieckhoff[32] also used a form of adiabatic magneti-zation to disturb the spin populations. They found that, by connecting a d.c. supply to the modulation coils of their magnet, they could change the value of their magnetic field by some 250 gauss in approximately 10 msec. This had the effect of altering the spin temperature, and hence of disturbing the population distribution, and had the advantage that it could be used to measure T_1 at values of magnetic field not confined to the resonance value at the microwave frequency. The disadvantage was that the change in population, and hence the rotation produced, was much less than could be obtained using the microwave pulses.

This technique has not been widely used. Whereas an error of less than 10 per cent has been claimed using concentrated crystals of $Nd(C_2H_5SO_4)_3 \ 9H_2O$ by paying careful attention to the design of the apparatus, it seems that the application of this technique may be rather limited. The rotation produced by many substances is small and, moreover, fairly large single crystals are required which must be of good optical quality and must be optically transparent. If the material is doubly refracting [as in the case of $Nd(C_2H_5SO_4)_3 \ 9H_2O$], it is exceedingly difficult to interpret measurements made in all directions other than the optic axis; this precludes any investigation of the variation of T_1 with crystal orientation. A further complication arises from the fact that Faraday rotation measures only the average of the spin populations and is thus subject to the limitations already discussed in Chapter

5 when more than two levels are involved. There may be cases where this technique will give additional information, for example concerning localized changes in spin temperature, and publication of further data is awaited with interest.

References

1. Feher, G., *Bell Syst. Techn. J.*, 1957, **26**, 449.
2. Assenheim, H. M., Introduction to Electron Spin Resonance (Adam Hilger Ltd, 1966; Plenum Press, New York, 1967).
3. Scott, P. L., and Jeffries, C. D., *Phys. Rev.*, 1962, **127**, 32.
4. Standley, K. J., and Wright, J. K., *Proc. Phys. Soc.*, 1964, **83**, 361.
5. Feldman, B. W., and McAvoy, B. R., *Rev. Sci. Inst.*, 1961, **32**, 74.
6. Bowers, K. D., and Mims, W. B., *Phys. Rev.*, 1959, **115**, 285.
7. Lees, R. A., Ph.D. thesis, Nottingham University, 1967.
8. Wilmshurst, T. H., *Electron Spin Resonance Spectrometers* (Adam Hilger Ltd, 1968; Plenum Press, New York, 1967).
9. Brown, G., Mason, D. R., and Thorp, J. S., *J. Sci. Inst.*, 1965, **42**, 648.
10. Gordon, J. P., *Rev. Sci. Inst.*, 1962, **32**, 658.
11. Taylor, P. F., Ph.D. thesis, St Andrews University, 1968.
12. Davis, C. F., Strandberg, M. P. W., and Kyhl, R. L., *Phys. Rev.*, 1958, **111**, 1268.
13. Pace, J. H., Sampson, D. F., and Thorp, J. S., *Proc. Phys. Soc.*, 1960, **76**, 697.
14. Giordmaine, J. A., Alsop, L. E., Nash, F. R., and Townes, C. H., *Phys. Rev.*, 1958, **109**, 302.
15. Baker, J. M., and Ford, N. C., *Phys. Rev.*, 1964, **136**, A1692.
16. Unruh, W. P., and Culvahouse, J. W., *Phys. Rev.*, 1963, **129**, 2441.
17. Solomon, P. R., *Phys. Rev.*, 1966, **152**, 452.
18. Taylor, A. G., Olsen, L. C., Brice, D. K., and Culvahouse, J. W., *Phys. Rev.*, 1966, **152**, section 1, 403.
19. Schultz, M. B., and Jeffries, C. D., *Phys. Rev.*, 1966, **149**, 270.
20. Standley, K. J., and Vaughan, R. A., *Phys. Rev.*, 1965, **139**, A1275.
21. Donoho, P. L., *Phys. Rev.*, 1964, **133**, A1080.
22. Standley, K. J., and Tooke, A. O., *J. Phys. Soc. C* (*Proc. Phys. Soc.*), 1968, **1**, 149.
23. Mims, W. B., Nassau, K. and McGee, J. D., *Phys. Rev.*, 1961, **123**, 2059.
24. Damon, R. W., *Cruft Lab. Tech. Rept.*, **136** (Harvard University, 1952).
25. Bloembergen, N., and Wang, S., *Phys. Rev.*, 1954, **93**, 72.
26. Bloembergen, N., and Feng, S. Y., *Phys. Rev.*, 1963, **130**, 531.
27. Lees, R. A., Moore, W. S., and Standley, K. J., *Proc. Phys. Soc.*, 1967, **91**, 105.
28. Feng, S. Y., *Cruft Lab. Tech. Rept.*, **371** (Harvard University, 1962).
29. Kastler, A., *Compt. rendu*, 1951, **232**, 453.
30. Opechowski, W., *Rev. Mod. Phys.*, 1953, **25**, 264.
31. Rieckhoff, K. E., and Griffiths, D. J., *Can. J. Phys.*, 1963, **41**, 33.
32. Daniels, J. M., and Rieckhoff, K. E., *Can. J. Phys.*, 1960, **38**, 604.

Chapter 8

Other Resonant Methods

§8.1

SPIN-ECHO TECHNIQUES

The techniques by which spin echoes are studied have not been widely used in the case of electronic spin systems, the names of Mims and his collaborators being perhaps most frequently associated with this work. When applied primarily to relaxation problems, electron-spin-echo data do not necessarily lead to parameters directly comparable with those obtained, for example, by pulse-saturation studies on similar materials; the parameters may give complementary information, but they may also be quite unconnected. As with other techniques, the choice of unsatisfactory experimental conditions may lead to results which are most difficult to interpret sensibly.

We shall first present the general features of the spin-echo experiment. The technique was first applied and analysed by Hahn[1] in 1950 and the reader is referred to his paper for the mathematical details of echo formation.

Homogeneous and inhomogeneous lines

If an assembly of isolated non-interacting identical spins is placed in a steady field H_0, one would expect all spins to precess about H_0 with a constant Larmor frequency. In a solid, however, there is a variety of perturbations which prevents this ideal condition from being a real one. As a result, there is a spread in the precessional frequencies which gives rise to the line width observed in resonance experiments. This line width may be considered to have two possible contributions, homogeneous and inhomogeneous. The homogeneous broadening is dynamic in character and occurs mainly through spin-spin interactions involving the same Larmor frequency, or through spin-lattice interactions. In effect, the field seen by a particular spin differs from the steady value H_0 by an amount which varies with time during an experimental

observation. The inhomogeneous broadening is essentially static
in origin and is due frequently to spatial inhomogeneities in the
applied magnetic field, to differing arrangements of magnetic
nuclei on neighbouring ions or of the magnetic moments of other,
not necessarily identical, paramagnetic ions or to imperfections in
the crystal sample used.

When inhomogeneous broadening is the major contributor to
the line width, one can speak of homogeneously-broadened spin
'packets', each packet having its own independent history of interactions with the magnetic fields. In particular, spin-lattice interactions normally cause spin flips in a random manner and thus give
rise to random variations in the local magnetic field at any neighbouring lattice site. Spin–spin interactions, causing spin–spin flips,
also introduce local field fluctuations and hence local variations
in the Larmor precessional frequency at neighbouring sites. The
frequencies of the individual spins in any of the homogeneously
broadened 'packets' are therefore varying with time in a complex
and unpredictable manner. However, if one considers the properties of the collection of spins which together make up a particular 'packet', certain properties emerge whose implications will
be considered below. In particular, one should note that, because
of the static nature of the inhomogeneous broadening, the spins
in one packet do not communicate readily with, and are not easily
affected by, the spins in another packet. As shown by Portis,[2] it is
thus possible in the detailed treatment of the behaviour of inhomogeneously-broadened lines to consider each packet as being
dynamically independent.

The spin-echo experiment

We shall give here an indication of the spin behaviour which
results in the formation of a spin echo, using a simplified model
first published by Hahn.[1]

In order to describe the effects clearly, it is advantageous to
consider the effects transformed to a co-ordinate system which
rotates at some convenient frequency. In the spin-echo experiment,
a steady field H_0 is applied along the z-direction and a pulse of
radio-frequency power, giving a radio-frequency magnetic field
H_1, is applied along the x-direction. The co-ordinate system is then
conveniently chosen to rotate about the z-axis with the same
angular frequency ω as the radio-frequency field. In the rotating
system, the radio-frequency field appears as a vector $\mathbf{H}_1$ in the
x'-direction, while H_0 along the z'-direction disappears except for
a residual component ΔH which is proportional to the difference
between the Larmor frequency of the particular spin packet and

ω, which we shall call $\Delta\omega$. When no radio-frequency field is applied, each spin packet precesses about its own ΔH, and when the radio-frequency pulse is on, precession takes place about $\mathbf{H} = \mathbf{H}_1 + \Delta\mathbf{H}$. If $H_1 \gg \Delta H$, the resultant field H lies almost along the x'-axis and has effectively the same magnitude for all the

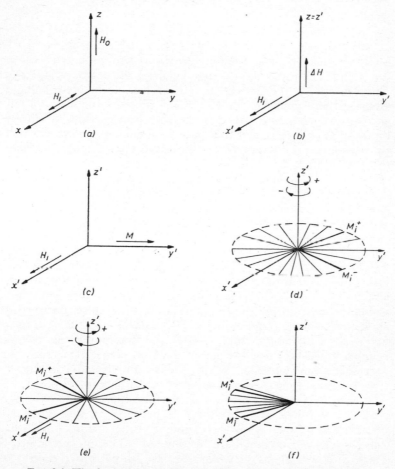

FIG. 8.1. The formation of a spin echo. The static co-ordinate system in (a) transforms to the rotating co-ordinate system; (b), rotation about the z-axis occurring at the Larmor frequency. Immediately following the $90°$ pulse, the magnetization vector is along y', (c). The individual vectors M_i then fan out as in (d) reducing M to zero; the sense of rotation of M_i^+ and M_i^- is shown. The $180°$ pulse rotates the vector disk of (d) about x', as in (e). At the time of the echo, the vectors M_i converge as in (f).

spin packets, which therefore all precess in the same manner during the radio-frequency pulse.

The sequence of events is readily followed with reference to Fig. 8.1. When $t = 0$, at the onset of the radio-frequency pulse, the magnetization of all the spin packets points along the z'-axis, the system being in thermal equilibrium. The pulse is applied for a time t_ω such that

$$\gamma H_1 t_\omega = \frac{\pi}{2}$$

and hence all the magnetization vectors precess through 90° about $\mathbf{H}_1$ from the z' direction, i.e. they lie along the y'-axis (Fig. 8.1c). After the pulse, the individual spin packets again precess about $\Delta \mathbf{H}$, i.e. they 'rotate' in the x'-y' plane. For the simplicity of our explanation we shall assume that the spin packets are isotropically distributed over the small and finite frequency range $\Delta \omega$ and that the time interval τ between the two radio-frequency pulses satisfies the condition $\tau \gg 1/\Delta \omega$. We can consider pairs of vectors $M_i{}^+$ and $M_i{}^-$, corresponding to frequencies $\pm \Delta \omega_i$, which precess in opposite directions about the positive z'-direction. In a time less than τ, an isotropic distribution of the vectors M_i will occur in the x'-y' plane and the resultant magnetization will have decayed to zero (Fig. 8.1d). At time τ, a second radio-frequency pulse is applied of length $2t_\omega$, so that $(2t_\omega)\gamma H_1 = \pi$. This time, each magnetization vector precesses through 180° about $\mathbf{H}_1$; the 180° pulse in effect turns over the magnetization disk of Fig. 8.1(d), about the x'-axis. After the pulse, the vectors precess about ΔH as before, but, during the time τ following the pulse, the vectors characterizing the faster spin packets catch up and the slow ones fall back until, at the time 2τ after the first pulse, the vectors occupy only a very small region of the x'-y' plane. A macroscopic precessing magnetization again occurs and a radio-frequency signal can be emitted into a suitably placed circuit, Fig. 8.1(f).

This is a simplified picture of the formation of an echo, but it is correct in its essential features. In practice, the pulses may be of such widths as to cause precessions by other than 90° and 180° and H_1 may not be much greater than ΔH. The interpretation of the echo experiments may then be less simple but the echo is inherently similar to that described above. In particular, it may be experimentally convenient to use two pulses each approximately $\pi/2$. In such a case it can be shown that the ratio between the static magnetization M_0 and the precessing magnetization M_p which generates the echo is given by

$$\frac{M_p}{M_0} = -\tfrac{1}{2} \sin \omega_1 t_\omega + \tfrac{1}{4} \sin 2\omega_1 t_\omega$$

where $\gamma H_1 = \omega_1$. The maximum value of this expression is theoretically 0·65 times that which the $\pi/2$, π pulse sequence might achieve (Mims[3]).

Although the above description is similar to that given by Hahn, with reference to the precession of nuclear spins at radio frequencies, the explanation holds adequately for the case of electron spins, provided the field H_1 is understood to be the microwave field in the resonant cavity, in the region of the paramagnetic sample.

Derivation of relaxation parameters from echo experiments

The above description of the echo experiment is idealized, particularly in the sense that it has been inferred that a magnetization vector M_i, tipped from the z-direction at $t = 0$ and subjected to a further rotation at $t = \tau$, remains throughout the sequence of events unchanged in the sense that its past history is well defined and 'remembered'. Clearly this will not be true in practice. Reference was made to the random spin flips caused by both spin-lattice and spin-spin interactions; the effect of these is to destroy the phase memory, which is an essential feature of the simple explanation given in Fig. 8.1. Thus, provided τ remains short, a spin echo should be observed, although its magnitude will have been reduced by the random spin flips. If τ is so long that all phase memory is destroyed, then clearly no echo will be seen at all. The maximum useful value of τ which still permits the observation of an echo depends, *inter alia*, on the spin-lattice and spin-spin relaxation times, T_1 and T_2 respectively. A study of the echo amplitude as a function of τ can therefore, under favourable circumstances, lead to information concerning T_1 and T_2.

A detailed theoretical study of these phenomena has been made by Klauder and Anderson.[4] The mathematical model used for the analysis is as follows. At any time t, there is a distribution of the spin frequencies in the whole sample given by $g(\omega)$, which is independent of t; this distribution occurs when the sample is placed in a uniform and constant magnetic field H_0. The spins can be divided into two groups, group A containing those spins which are actually observed and whose frequency distribution is given by $P_1(\omega)$, where the range of $P_1(\omega)$ is essentially very much smaller than $g(\omega)$, and group B containing spins with frequencies within $g(\omega)$ but outside $P_1(\omega)$. (The notation is that of Klauder and Anderson.) The B group will generally contain many more spins than the A group and, during the period of echo formation and observation, spin flips within the A group are ignored. However, reorientation of certain spins within the B group will take place on account of both T_1 and T_2 processes. Thus the local fields

L

acting on A spins will alter, with resulting changes in the local fields and hence in the local Larmor frequencies of these spins. A general form for the spectral-diffusion law is derived

$$P(\omega, t; \omega_0) = (2\pi)^{-1} \int \exp[iy(\omega - \omega_0) - t f(y)] \, dy$$

where $P(\omega, t; \omega_0)$ is the probability of finding a frequency ω after diffusion has occurred for a time t, from an initial frequency at ω_0. Two forms of the parameter $f(y)$ are discussed, $f(y) = ky^2$ and $f(y) = m|y|$, leading to decays of the two-pulse echo of the form

$$\frac{A_\tau}{A_0} = \exp(-2k\tau^3/3) \quad \text{when } f(y) = ky^2 \qquad (8.1a)$$

and

$$\frac{A_\tau}{A_0} = \exp(-m\tau^2) \quad \text{when } f(y) = m|y| \qquad (8.1b)$$

The first of these represents a Gaussian type of diffusion, and the second a Lorentzian one. Here, A_τ is the amplitude of the echo when there is a time τ between pulses.

Mims *et al.*[5] found the Lorentzian type of line to be a better approximation to the observed behaviour in some cases of cerium and erbium ions lightly doped into calcium tungstate crystals, but certain deviations from the Lorentzian model were found.

In certain favourable cases, when the parameter m of equation (8.1b) had been experimentally determined, T_1 could be found. Klauder and Anderson give the theoretical expression for m as

$$m = \tfrac{2}{3}\pi^2 \mu \gamma n r \qquad (8.2)$$

where μ is the moment of a single B spin and γ is the gyromagnetic ratio. The density of the B spins is given by n, and the average rate at which spins flip is

$$r = \frac{1}{T_2} + \frac{1}{T_1} \qquad (8.3)$$

Thus T_1 can be found when the other parameters are known; it may happen that $1/T_2$ is negligible compared with $1/T_1$.

Mims *et al.* found that m was not always easily derived from two-pulse data. They used also three-pulse sequences, with a variable time T' between the second and third pulses, a 'stimulated echo' being emitted after a time $2\tau + T'$. They again measured A_t/A_0 and found

$$\frac{A_t}{A_0} \propto \exp[-T'f(\tau)]$$

in some cases where both

$$f(\tau) = m(\tau - \rho)$$

and

$$f(\tau) = m[(\tau^2 + \rho^2)^{\frac{1}{2}} - \rho]$$

fitted the experimental data equally well. Klauder and Anderson derived the second form of $f(\tau)$ theoretically with

$$\rho = \pi R_{min}^3 / 2\mu\gamma$$

where R_{min} is the nearest approach of a B spin to an A spin.

Thus the spin-echo experiment, carried out over a range of values of τ, leads to an experimental determination of the parameter m or k. In the work of Mims *et al.* the value of m was 3×10^9 sec^{-2}, for example, for a sample of $CaWO_4$ containing 0.83×10^{18} Ce^{3+} ions/cc and 0.99×10^{18} Er^{3+} ions/cc. The temperature of measurement was $1.6°K$ and the magnetic field was applied at $90°$ to the c-axis of the crystal. The experimental value of ρ was 2 μsec.

Clearly, in order to derive further information, one must know which, if either, of the processes control the relaxation rate r. One reasonably reliable indication is the variation of the calculated value of r with temperature—if 'T_2' processes predominate one expects r to be insensitive to changes in temperature while, in the helium temperature range, 'T_1' processes are likely to follow something approaching the T^{-1} law of the Van Vleck direct process. This point appears to be one of some difficulty in the general application of the spin-echo technique to relaxation-time measurement in that there must be some uncertainty in deciding *a priori* whether a particular ion in a given lattice is going to yield experimental data capable of giving reliable T_1 values. Referring to the work of Mims *et al.*, Klauder and Anderson calculated m from equation (8.2), having observed from the temperature variation that this might be considered a good 'T_1' sample. The data used were $g_{Er} = 8$, $g_{Ce} = 1.4$ and $r = 1/T_1 = 10^2$ sec^{-1}. This latter value was determined experimentally by a pulse-saturation technique. The calculated value of m was 1.7×10^8, a factor of 20 less than the observed value. The most obvious source of error is the experimental value of T_1. Echo measurements are completed within 1 msec, and this time is quite insufficient for 10-msec T_1 processes to be important. Thus it appears likely that echo diffusion studies essentially relate to T_1 processes which occur only in the very early stages of a pulse-saturation/recovery experiment. The indication in the Mims data is that these processes are faster by at least an order of magnitude than those governing the later recovery. This is the justification for the earlier statement of the complementary nature of spin echo and pulse-saturation data.

Mims[3] has studied the sensitivity of electron-spin-echo methods in normal E.S.R. spectrometry. He comes to the conclusion that the ultimate sensitivities of both CW and spin-echo methods of spectrometry are of the same order of magnitude provided the two instruments operate at the same average power level and with the same time of observation. Mims notes certain experimental advantages of the echo method; for instance there is an unambiguous base line and there is no need for any microwave bridge. The elimination of the latter removes the limit on sensitivity which is sometimes set by mechanical vibrations and by slight frequency variations. Noise of this type modulates an existing signal and, in the echo case, does not determine whether or not such a signal is seen. He makes the point that the echo method might possess decisive advantages if one were studying very broad resonance lines, partly because of the reliability of the base line, but particularly because long and relatively flat portions of the line would still contribute full-strength signals. A serious limit on the echo method is the requirement that the phase-memory time should be four or five times as great as the pulse duration. So far this requirement has limited measurements to materials with phase-memory times of the order of 1 μsec. These points apply, of course, only to the use of the echo method in relaxation measurement. There is, however, an additional experimental advantage which may in certain circumstances be very important. In the pulse-saturation technique one has to pay particular attention to keeping the level of the monitor power well below that which might cause any saturation of the line being studied. For instance, when working with the long-relaxation-time ruby materials, the authors have operated with monitor powers well below 1 μW. This necessarily causes a limit to be set to the sensitivity of the detecting spectrometer. Such a limitation does not exist when one is using the echo technique, for one hopes to arrange for as many of the spins as possible to be affected by the first pulse and, ideally, all these spins contribute to the second pulse which is radiated into the cavity when there is no microwave signal present.

Experimental techniques

The apparatus required for spin-echo work is basically similar to that used in the pulse-saturation technique, but with the addition of a time-controlled pulsing sequence. There is the further important requirement that the detecting and amplifying stages shall have a very short recovery time since the measurements of the echo usually take place within microseconds of the second or final pulse. Thus some additions must be made to the circuits in order to give adequate protection against saturation.

Fig. 8.2 is a block diagram of the apparatus used by Mims.[3] Two microwave pulses each of 0·2 μsec duration and, in this case, up to 50 W peak power, are generated by a Varian V63 klystron at a frequency of 9·42 Gc/s. A primer klystron is employed providing a weak signal in the V63 cavity to eliminate time jitter which was found to be present when the V63 was required to build up to full power from thermal noise. The primer was tuned 10-20 Mc/s away from the V63 centre frequency, and at this setting its frequency was outside the acceptance band of the cavity and the detection circuits. The high-power pulses from the V63 were fed through a variable attenuator and directive feed to a tunable rectangular reflection cavity operating in a H_{011} mode with a loaded Q of about 1500. The microwave field amplitude H_1 in the

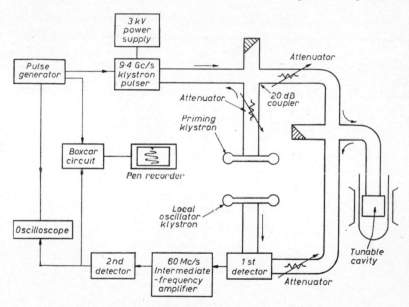

FIG. 8.2. Block diagram of the spin-echo apparatus used by Mims.[3]

cavity was of the order of 0·5 gauss and the actual microwave peak pulse power required was not more than 1 W. In Mims' case the echo signal was fed from the cavity through a variable attenuator to a balanced mixer and thence to a superheterodyne system with an intermediate frequency of 60 Mc/s and a band width of 8 Mc/s. From the second detector, the signal went simultaneously to an oscilloscope and to the input of a 'boxcar' integrator circuit. The function of the latter was to integrate a number of echo signals, and the output of this was fed to a pen recorder. For the amplitude measurements required when determining T_1 the oscilloscope was used. This is essentially the spin-echo apparatus used for spectrometry. In the work described by Mims, Nassau and McGee[5] pulses of microwave power at 6·7 Gc/s were provided of 1 μsec duration by pulsing a grid-controlled travelling-wave-tube amplifier. The detection system in this case had an overall response time of 0·25 μsec and contained a paralysing circuit to prevent overloading of the later stages by the power pulse.

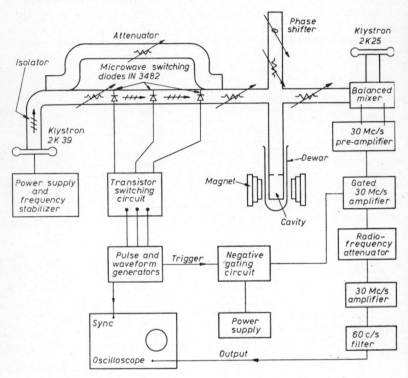

FIG. 8.3. Block diagram of spin-echo apparatus (Dyment[6]).

Dyment[6] used a somewhat different system which is shown in Fig. 8.3. He used three IN3482 microwave switching diodes in the microwave circuits between the klystron and the cavity, and the circuits operating these diodes produced the necessary pulsing. Pulse durations were fixed at 1 μsec with a repetition interval of 8 msec. The power from the 2K39 klystron was thus varied between 3 and 59 dB in this way.

Breen[7] adopted yet a third slight variation in order to produce the necessary protection of the circuits. A block diagram is shown in Fig. 8.4. Breen's pulses were generated by Tektronix waveform and pulse generators T162 and T163, and the pulses had a rise and fall time of approximately 30 nsec. One pulse sequence was applied to the English Electric K350 klystron giving peak powers of the order of 2 W. Another, made slightly longer for protection, was similarly applied to the low-power local oscillator klystron. Thus there was no signal passed to the amplifying system during the high-power pulse. Sufficient protection for the microwave detecting diodes was provided by a pair of crossed 1S44 diodes incorporated across each detecting arm. These diodes began to conduct when potentials of about 0·4 V were built up across them so that higher voltages could not be applied to the detecting diodes.

The decay envelopes for the two-pulse echoes are obtained by varying the time τ between the two pulses and superimposing a large number of traces on an oscilloscope. In a stimulated echo

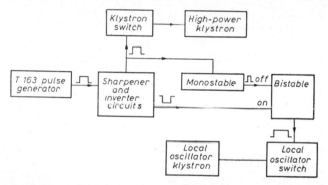

Fig. 8.4. Block diagram of spin-echo pulse generation and protection system used by Breen.[7]

experiment, τ is kept constant and T is varied. In both cases, the time intervals may be measured sufficiently accurately by relying on the calibrated time base of a reliable oscilloscope. There is clearly a requirement that a trace length of the order of 1 μsec shall be large and well defined, but with modern oscilloscopes this does not present any real difficulty.

§8.2

THE INVERSION-RECOVERY TECHNIQUE

In §7.1 the experimental method was described in which a transition is saturated by a high-power pulse and the return of the populations to the thermal equilibrium distribution is monitored.

In that method two energy levels were at best saturated, that is to say, their populations were made equal. In this section we shall describe a logical development of that method wherein the levels with their original populations are inverted, and the recovery of the original thermal equilibrium is again observed. The inversion of the two levels is usually achieved in one of two ways, by means of a 180° pulse, or using the technique known as 'rapid passage'.

The inversion techniques

The 180° pulse and its effects are discussed fully in §8.1. The name was first given by Hahn to a pulse of radio-frequency power of sufficient intensity and applied for a time $2t\omega$ such that

$$2t\gamma^\omega H_1 = \pi$$

As a result a spin, initially precessing about the static Zeeman field H_0 which is applied along the z-axis, is completely turned over so that at the end of the pulse it is precessing about the $-z$-axis.

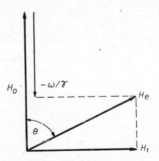

FIG. 8.5. The effective field H_e seen by the spin system in the rotating co-ordinate system. H_e is the vector sum of H_0, H_1 and $(-\omega/\gamma)$.

The technique of adiabatic rapid passage was described by Bloch[8] and by Abragam.[9] Consider an assembly of spins when it is viewed from a rotating co-ordinate system (Fig. 8.5). The spin is influenced by three magnetic fields, (*a*) the Zeeman field H_0 directed along the z-axis (which coincides with the z'-axis), (*b*) the radio-frequency field H_1 (which appears as a steady field along the x'-axis in the rotating framework) and (*c*) a field $-\omega/\gamma$ opposing H_0, which is produced or introduced by the rotation of the co-ordinate system at the frequency ω.

The spin system must precess about the resultant field direction H_e and produces a net magnetization vector **M** along this direction also.

The name 'adiabatic passage' through the resonant condition $(\omega = \omega_0 = \gamma H_0)$ is given to a slow variation of the frequency ω from a value much below ω_0 to a value well above this. In this context, a slow variation means slow in comparison with the Larmor period. In Fig. 8.6(a), viewed from the rotating framework, ω is very much less than γH_0, so that H_e lies close to the direction of H_0. At resonance when $\omega = \omega_0$ (Fig. 8.6b), $\omega_0 = \gamma H_0$, and the resultant field is just H_1 in the x' direction. When ω is very much greater than γH_0 (Fig. 8.6c), the net field H_e is due to H_1 along x' and $[H_0 - (\omega/\gamma)]$ along the $-z'$ direction; H_e lies close

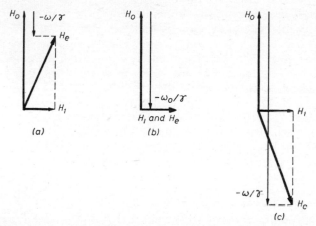

Fig. 8.6. Directions of H_e as the frequency ω is varied from a value much below the resonant frequency ω_0 to a value much above it, viewed from a rotating framework. (a) $\omega \ll \gamma H_0$, (b) $\omega_0 = \gamma H_0$, (c) $\omega \gg \gamma H_0$.

to the $-z$ direction if H_0 is very much greater than H. The adiabatic passage has therefore rotated the effective field from its original z' direction through x' and into the $-z'$ direction; the direction of the magnetization has followed the same route, that is to say it has been reversed in direction.

In practice one usually keeps ω constant at a value ω_0 and sweeps the Zeeman field H slowly through the value H_0. At the end of the sweep, the two methods have both had the same result, namely a net magnetization vector has been produced which is directed opposite to the applied Zeeman field H_0. If we consider the case of two spin levels only and label the levels 1 and 2, then the population n_1 in the lower level is greater than n_2 in the upper level. The magnetization at normal temperatures in the direction of the applied field is proportional to $n_1 - n_2$ (see §7.2). As a result of the inversion experiment described above, the magnetization vector is reversed in direction but unaltered in length. This occurs if the n_2 spins in level 2, which were previously in the higher energy state, now find themselves with lower energy than the n_1 spins in level 1. Levels 1 and 2 have been interchanged without altering the populations in the two levels. There is now an excess of spins in the upper levels, and if the spin temperature is defined by equation (1.11), p. 10, we have a system which corresponds to a negative temperature.

When the radio-frequency field H_1 is removed, at the end of the 180° pulse for example, the only field remaining is H_0, and the n_1 spins in the (temporarily) upper level are now in an unfavourable and unstable energetic condition. They must lose energy and regain their former position of lower energy. Another description of this state of affairs is that when H_1 is removed, the equilibrium position of the magnetization vector is directed along H_0 rather than against it and therefore the vector will tend to swing round to its previous position. Whichever way one describes it there must be interaction with the lattice for this to occur.

In both the experiments described, the radio-frequency field is obtained from a pulse of microwave power; in the second method it is applied during the time that the field is swept through the resonance condition.

Outline of the experimental methods used

A 180° pulse was used by Collins, Kyhl and Strandberg[10] to achieve inversion in crystals of potassium chromicyanide containing 0·1 per cent Cr^{3+}. They obtained the required 10 nsec pulses in a novel way, from the energy stored in a waveguide cavity which was excited by means of $\frac{1}{2}$ μsec magnetron pulses (see Fig. 8.7). One end of the waveguide could be effectively opened by firing a spark; this unbalanced a magic tee bridge by changing the plane at which reflection occurred. The sample was placed in a cavity possessing two degenerate orthogonal modes. The short pulse was applied in one mode where there was only a low Q of the order of 30, and magnetic resonance was observed, in the usual manner, using the other mode which had a high Q-factor.

The rapid passage method has been rather more widely used in E.S.R. relaxation work. Castle, Chester and Wagner,[11] and Rannestad and Wagner[12] used the timing sequence shown in Fig. 8.8. The steady magnetic field was held slightly below the resonant value. A current pulse of the form shown in line (*a*) was applied to a pair of Helmholtz coils mounted on the cavity which was immersed in liquid helium. As a result, the field was swept through resonance twice and the absorption line was thus seen twice at the output of the spectrometer, as shown in line (*b*). For rapid passage work, a pulse of microwave power was applied at the resonant frequency of the cavity during the first sweep (see *c*). Inversion of the line resulted and was observed on the downward or returning sweep of the fields, as shown in (*d*). However, a later

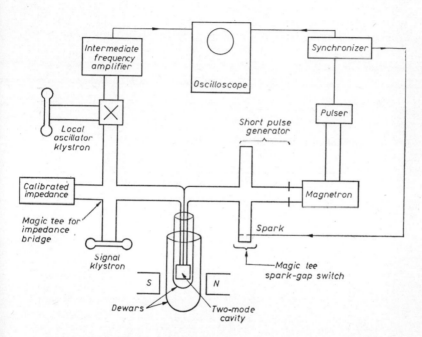

FIG. 8.7. Block diagram of the equipment used by Collins *et al.*[10] to produce the 180° microwave pulse required to invert a spin system. The energy stored in the waveguide cavity may be released by firing the magic tee spark-gap switch.

application of further field sweeps without the pulse permitted observation of the whole line during its recovery as (*d*) shows. The relaxation time constant may be obtained from the envelope of these absorption lines plotted as a function of time. The quoted conditions for rapid passage were: microwave field strength 0·5 to 2·0 gauss, time of passage through the line of the order 10^{-4} seconds with T_1 of the order of 10^{-2} seconds. Sweeps of up to 120 gauss in times not less than 20 μsec were obtained from the Helmholtz coils. This necessitated using a cavity constructed from silvered epoxy resin in order that the swept field could penetrate the walls of the cavity without attenuation.

The above method whereby fast passage is induced and electron spin resonance is measured at the same frequency is of necessity a comparatively slow one. This must be so when one considers the type of field variation; dH/dt must change sign. In order to overcome this, Brya and Wagner[13] used two Zeeman frequencies, one for fast passage v_p and one for measurement v_m. A silver-plated plastic bimodal cavity was used operating in the H_{102} and H_{011} modes at frequencies of 11·35 and 11·45 Gc/s respectively. The

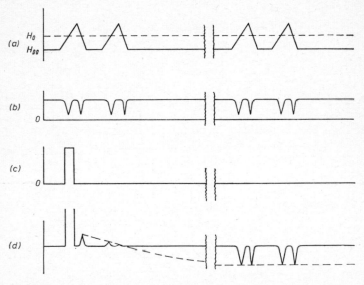

FIG. 8.8. Timing sequence used by Rannestad and Wagner[12] for fast-passage recovery measurements. (a) Field produced by a current pulse applied by Helmholtz coils, (b) spectrometer output display of absorption lines, (c) microwave pulse, and (d) recovery of inverted line.

magnetic field could therefore be swept continuously through the two resonant conditions with very little time delay. It was also relatively easy to prevent saturation of the receiver by the high-power microwave pulse. The pulse required for fast passage was at a frequency v_p whereas the receiver was tuned to a different microwave frequency v_m and rejection of v_p was not difficult to arrange. Pulses were obtained from a travelling-wave amplifier giving peak powers of 50 W and pulse durations of the order of 5 μsec. Variable coupling to each cavity mode was achieved by using two loops which could be adjusted from outside the helium cryostat. In order to make continuous measurements of the relaxation of the centre of the resonance line (as is the requirement in the pulse-saturation method) inversion was induced on the up sweep of the magnetic field at a field value H_p corresponding to the resonant frequency v_p of one cavity mode, and measurements were possible when the field reached its steady value H_m corresponding to the resonance value at v_m. The decay of the centre of the line could then be observed, recorded and analysed as described in Chapter 7.

In order to display the entire resonance line during recovery, the field was swept continuously through both H_p and H_m with a variable delay between successive field sweeps.

The relative merits of this technique

The major disadvantages of this method are the stringent requirements of a controlled field sweep and a short synchronized high-power microwave pulse. On the other hand there are several advantages. Inhomogeneous broadening of resonance lines can be detected immediately by means of the hole burning which appears in the line and, even if this is present, it does not introduce complications of interpretation. In the pulse-saturation method the spin diffusion following the hole burning would be a serious complicating factor, but this is not so here as microwave energy is supplied across the whole line. A relatively high monitoring power can also be tolerated without disturbing the line, if the recovery of the complete line is observed.

By far the most important advantage is the sensitive test provided for the presence of a phonon bottleneck. We are here dealing with the interaction between the spin system and the lattice, that is to say, interaction between a system initially at a negative temperature with one whose temperature must always be positive. The excited spin system will endeavour to return to thermal equilibrium with its surroundings by means of absorption and emission of phonons. Since, when inversion is complete, there are more spins in the upper level than in the lower level, then more phonons are emitted than are absorbed. It is possible for the emitted phonons to be immediately absorbed by the phonon bath when the decay will proceed exponentially. If, however, for some reason these phonons are not immediately absorbed, they may interact with the spin system and stimulate further emission before they reach the crystal boundary to be absorbed finally by the bath. Clearly the relaxation will be accelerated in a regenerative manner and there results what can be described as a phonon avalanche. However, when the phonons and spins reach a common positive temperature, the avalanche can no longer occur and phonon imprisonment sets in, the recovery now being retarded. Thus, in the presence of a phonon bottleneck, the observed relaxation parameter is first markedly decreased and then much increased; the change in its magnitude is discontinuous and occurs when the number of spins in the upper level is instantaneously equal to the number of spins in the lower. Thus a plot of the course of the recovery shows two distinct parts, with an abrupt break at the point corresponding to

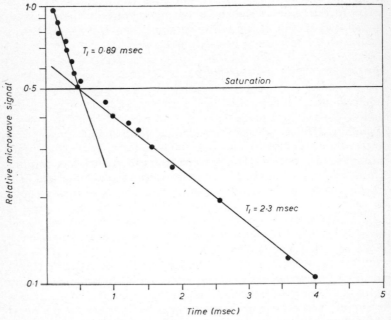

FIG. 8.9. Discontinuity in recovery trace obtained by Firth and Bijl[14] for F centres in irradiated MgO.

complete saturation, i.e. to an infinite spin temperature. When there is no bottleneck the recovery has an exponential form which is continuous through the point of saturation. Thus the presence of the bottleneck is shown by the sharp discontinuity and there can usually be no doubt about the interpretation.

A good example of such behaviour is shown in Fig. 8.9. These results were obtained by Firth and Bijl[14] for F centres in irradiated MgO and such discontinuities were observed only in their more concentrated samples ($> 10^{18}$ spins). Brya and Wagner[13] have also observed this type of relaxation behaviour in 0.5 per cent cerium-doped lanthanum magnesium nitrate.

§8.3

SOME OPTICAL TECHNIQUES

(a) Excited states:

Information obtained from electron spin resonance and from the types of relaxation measurements already described earlier relate almost entirely to the ground-state energy levels of the paramagnetic system. Only in certain cases (for example when an Orbach relaxation mechanism is operative) can one obtain even indirect information about excited levels. It would be very desirable to study excited levels with the high resolution of E.S.R. techniques, but in the case of magnetic impurity ions in crystals the number of ions maintained in these states by methods such as optical pumping is well below the limiting sensitivity of conventional E.S.R. techniques. It is therefore not possible to monitor directly the absorption of microwave power by the excited states, but, if these states are fluorescent, then one may be able to monitor the change in the intensity of the fluorescent light from such levels when the microwave energy coincides with its Zeeman splitting. The increase in sensitivity thus obtained may be several orders of magnitude, since optical rather than microwave photons are now being detected. As few as 10^7 ions in the excited state can produce a detectable signal.

Geschwind *et al.*[15] review several optical methods of detecting microwave absorption. The most useful technique, and the one which they found most amenable to relaxation measurements, made use of a high-resolution optical scanning spectrometer to monitor the intensity of the individual fluorescent transitions. Consider the simple situation, depicted in Fig. 8.10, consisting of two doublet levels each split by a Zeeman field. By means of some continuous optical pumping mechanism, the upper doublet population is maintained in a steady state. Fluorescent transitions A and B are selectively monitored by means of a spectrometer, such as that shown in outline in Fig. 8.11. If a microwave transition is induced within the upper doublet, then the intensity of transition A will decrease, and that of B will increase.

Geschwind *et al.* investigated, in particular, the hyperfine structure of the $\bar{E}(^2E)$ state of the Cr^{3+} ion in ruby. Those parts of the energy-level scheme for this ion which are relevant to the following discussion are shown in Fig. 8.12. The main absorption bands in ruby are between the ground $^4A_2$ orbital singlet and the $^4T_2$ and $^4T_1$ triplets; these transitions are shown on the left of Fig. 8.12. Subsequent decay of these to the ^2E levels by radiation-

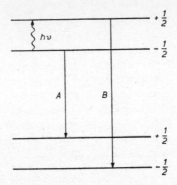

FIG. 8.10. Simple energy-level scheme for optical detection of E.S.R. Both the excited-state doublet and the ground-state doublet to which it fluoresces are split by a Zeeman field.

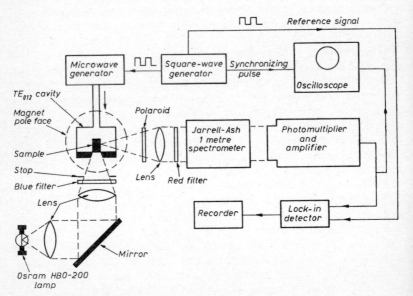

FIG. 8.11. Block diagram of the spectrometer used by Geschwind *et al.*[15] for the optical detection of E.S.R.

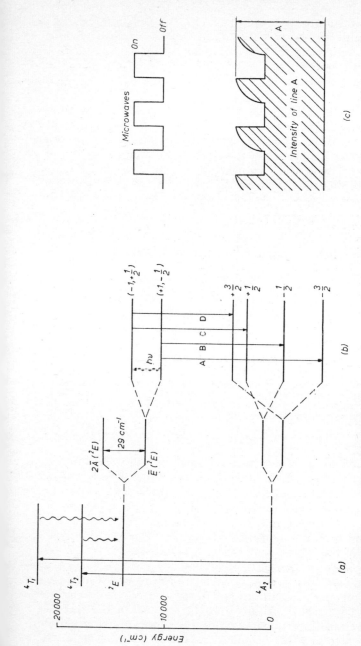

Fig. 8.12. Schematic energy level diagram for Cr^{3+} in Al_2O_3. In (a) are shown some of the levels in a cubic field. These are further split in (b) by higher-order perturbations and a magnetic Zeeman field. The allowed fluorescent transitions between the excited doublet and the ground states are shown. If a microwave field resonant with the excited level splitting is square-wave modulated, then the intensity of transition A will vary as shown in (c), an initial decrease, followed by a recovery due to spin-lattice relaxation and radiation to a ground state. The sample is continuously irradiated with mercury light.

less transitions are also shown in Fig. 8.12(a). The ^2E state is further split by higher order perturbations into two Kramer's doublets as shown in (b), and it is in the lower of these, the $\bar{E}(^2E)$, that electron spin resonance is performed. Some of the permitted fluorescent transitions between the $\bar{E}(^2E)$ levels and the ground $^4A_2$ levels are shown.

A steady-state population is maintained in $\bar{E}$ by a continuous excitation to the absorption bands with mercury light. At fixed magnetic field, corresponding to resonance in $\bar{E}$ for the microwave frequency used, a particular transition is monitored [e.g. line A in Fig. 8.12(b)] with the high-resolution optical spectrometer and photomultiplier tube. When the microwaves are switched on, the

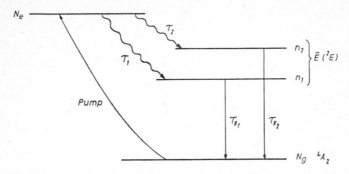

FIG. 8.13. Simplified diagram of the pumping cycle and recovery transitions. Levels 1 and 2 are filled by relaxation from an excited level whose population N_e is maintained by continuous optical pumping. τ_{R_1} and τ_{R_2} are the radiative lifetimes of levels 1 and 2 to the ground state.

intensity of line A is reduced, corresponding to a transfer of ions from the $(+1, -\frac{1}{2})$ level to the $(-1, +\frac{1}{2})$ level. When the microwaves are switched off, the $(+1, -\frac{1}{2})$ level will recover to its initial population and the intensity of line A will correspondingly increase [see Fig. 8.12(c)]. The monitored recovery time will depend on both the spin-lattice relaxation time *and* the radiative lifetime (of the excited state) to the ground state.

A simplified version of the pumping cycle and recovery transitions is shown in Fig. 8.13. In this are shown only those aspects of the foregoing mechanisms which are necessary to describe the recovery of the A transition after the removal of the microwave pulse. Even with the maximum possible continuous optical pumping, such factors as the magnitude of the optical absorption factor and the lifetime of the excited states mean that the population

$(n_1 + n_2)$ of the excited state is still orders of magnitude smaller than N_g the population of the ground state.

τ_{R_1} and τ_{R_2} are essentially the effective radiative lifetimes of levels 1 and 2 to the ground state and are usually unequal. They are also temperature-dependent in the region of interest ($1 \cdot 5 - 4 \cdot 2°K$), reflecting the redistribution of population among the $^4A_2$ ground-state levels with temperature. In addition, τ_{R_1} and τ_{R_2} will depend on concentration, geometry and method of viewing the light, and whether one is observing the peak or side of the optical line. Therefore τ_{R_1} and τ_{R_2} must be measured *in situ* under the identical conditions used in the microwave experiment. This may be done in a separate experiment using a pulsed light source.

n_1 and n_2 are the populations of the lower and upper Zeeman components of $\bar{E}$, and T_1 is the spin-lattice relaxation between these levels. If $\tau_{R_1} = \tau_{R_2} = \tau_R$, then $(n_1 + n_2)$ is approximately constant, independent of T_1, at equilibrium. In this case, the populations n_1 and n_2, after being disturbed by a microwave pulse, show an exponential recovery with a time constant τ_{obs} given by

$$\frac{1}{\tau_{obs}} = \frac{1}{\tau_R} + \frac{1}{T_1} \qquad (8.4)$$

If $\tau_{R_1} \neq \tau_{R_2}$, then $(n_1 + n_2)$ does depend on T_1 and in addition will not be constant during the recovery of the populations n_1 and n_2 towards equilibrium. Geschwind *et al.* found that equation (8.4) adequately described the recovery of the light signal from all their ruby samples at temperatures above $3 \cdot 3°K$, and at lower temperatures for the more dilute crystals. However, at low temperatures for their higher concentration samples, where $T_1 \sim \tau_R$, it was found necessary to employ the complete equations (derived in Geschwind *et al.*[15]) of which equation (8.4) is a simplification. At the lower temperatures, because of the increase in T_1, a greater degree of saturation was achieved and the change in intensity of the optical transition was as much as 10 per cent of the detected signal. At higher temperatures, however, averaging techniques had to be used (see for example p. 115) in order to extract the signal from the background noise (Fig. 8.14).

At low temperatures, Geschwind *et al.* found that the spin-lattice relaxation time T_1 was masked by the radiative lifetime τ_R when these became comparable in magnitude ($\tau_R \sim 6$ msec). A semi-log plot of T_1 for the $\bar{E}$ level against the reciprocal of the temperature (Fig. 8.15) yielded a straight line indicative of an Orbach relaxation process from which

$$T_1 \propto \exp(\Delta/kT)$$

with $\Delta \sim 29$ cm^{-1} in very good agreement with the splitting

between the ^2A and $\bar{E}$ levels (Fig. 8.12*b*) measured spectroscopically.

Addé *et al.*[16] have performed further measurements on the
same transitions in ruby, using the above technique, and have
found clear evidence of a phonon bottleneck limiting the recovery

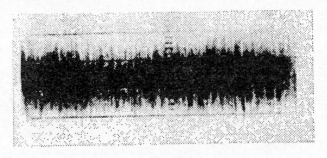

(a) Input to multichannel analyser.

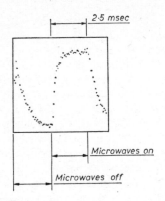

(b) Output (approx. 8 minutes running time).

FIG. 8.14. Example of the extraction of a modulated light signal previously buried in noise using 96 000 samplings. Theoretical improvement in signal-to-noise ratio is $(96\,000)^{1/2} \sim 305$ (Geschwind *et al.*[15]).

at low temperatures. It would appear that this method affords a
very convenient way of checking the concentration dependence
of T_1 which would be expected in such a situation. The populations n_1 and n_2 in the excited states may easily be altered over a wide
range simply by varying the intensity of the pumping light,
whereas in ground-state experiments one has to correlate results

obtained using several different samples, an operation which appears to be very suspect.

At low concentrations in the $\bar{E}$ state, the above authors found no evidence of any bottleneck effects, but using ruby samples of concentration 5×10^{18} per cc and 8×10^{18} per cc, at $4 \cdot 7°$K and at frequencies of 24 and 48 Gc/s, a marked dependence of the

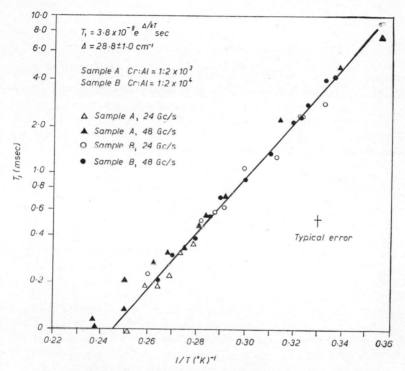

$$T_1 = 3 \cdot 8 \times 10^{-9} e^{\Delta/kT} \text{ sec}$$
$$\Delta = 28 \cdot 8 \pm 1 \cdot 0 \text{ cm}^{-1}$$

Sample A Cr:Al = 1:2 × 10³
Sample B Cr:Al = 1:2 × 10⁴

△ Sample A, 24 Gc/s
▲ Sample A, 48 Gc/s
○ Sample B, 24 Gc/s
● Sample B, 48 Gc/s

Typical error

T_1 (msec)

$1/T \, (°K)^{-1}$

FIG. 8.15. Typical results for Orbach relaxation in the excited $\bar{E}$ state of ruby (Geschwind *et al.*[15]).

measured relaxation time on pumping light intensity was observed. Even at the highest light intensities the recovery always exhibited a single exponential over a time equal to at least five times the observed T_1. The values of T_1 obtained under these conditions varied from about 71 ± 2 μsec at low light intensities (no bottle-neck) to about 300 μsec at the highest light intensity, corresponding to severe bottleneck conditions.

These optical measurements therefore have several advantages over the usual type of ground-state experiments. Not only does the above technique have the advantage of greatly increased sensitivity, but it also eliminates the effects of spin diffusion associated with hole-burning effects in inhomogeneously-broadened lines because the total population of the level is monitored by the rather coarse optical technique. Its main advantages seem to be the direct evidence it can furnish about excited states, and also the simple means of observing the concentration effects associated with a phonon bottleneck. It appears that most Orbach relaxations so far observed in excited states show evidence of a phonon bottleneck; Addé *et al.* suggest that the scatter of relaxation times reported by different authors for Orbach mechanisms could be due to the varying concentrations of their samples.

(b) Ground-state spin memory in optical pumping:

The fact that electron spin resonance can be observed in excited states (implying that there is a population difference between these excited Zeeman levels), in spite of the spin-lattice relaxation between these levels being much slower than the radiative decay, implies that there is selective pumping to these levels from the ground state. At very low temperatures the levels cannot come into thermal equilibrium before they radiate and therefore must have been preferentially populated. However, this was not considered by Imbusch and Geschwind[17] to be sufficient evidence that ground-state spin memory had been retained under broad-band optical pumping conditions, because such preferential pumping could also come about by thermalization in higher levels which feed the $\bar{E}$ state. In order to confirm this idea, these authors attempted to observe a change in the relative populations of the $\bar{E}$ levels when the populations of the ground $^4A_2$ spin levels had been readjusted by microwave saturation in the following way. With fixed microwave frequency, the magnetic field was swept through a ground-state resonance saturating it, and the change in light intensity observed using the same experimental arrangement as above. It was observed that when for example the $(-\frac{3}{2}) - (-\frac{1}{2})$ ground-state transition was saturated, the intensity of the A line (Fig. 8.12b) decreased whereas that of the D line increased, indicating an increase in the population of the $(-1, +\frac{1}{2})$ level relative to the $(+1, -\frac{1}{2})$ level. This change in population is expected from consideration of the simple selection rules $\Delta m_s = \pm 1, 0$. The observed temperature dependence of the population difference thus measured, in both dilute and concentrated ruby specimens,* would seem to confirm that spin memory is retained.

* In heavily-doped ruby, reabsorption of the A line is not negligible, and saturation of the $(-\frac{3}{2}) - (-\frac{1}{2})$ ground-state transition reduces the population in the $-\frac{3}{2}$ level, thus reducing the trapping of the A line and increasing its intensity. This effect opposes the spin-memory effect and correction for it is required.

*(c) Observation of a microwave phonon bottleneck by Brillouin
    scattering:*

We saw in Chapter 4 that, in the presence of a phonon bottleneck, the narrow band of phonons on speaking terms with the
spins is heated up to a temperature greater than that of the bath.
The pulse-saturation method of measurement (§7.1) gives indirect
evidence for such a situation, and the inversion-recovery method
(§8.2) would appear to furnish more direct evidence. Geschwind[18]
has proposed a method of looking directly and selectively at the
bottlenecked phonons themselves by means of Brillouin scattering.

The thermal fluctuations in a medium give rise to a locally
fluctuating dielectric constant which will in turn scatter a light
beam in all directions and impose upon it a frequency modulation
which reflects the spectral composition of the thermal fluctuations.
Pressure fluctuations propagate as acoustic waves and the light will
experience a Doppler shift when scattered by these phonons—this
is referred to as Brillouin scattering.

An acoustic wave will set up periodic regions of rarefaction and
compression in a medium so that the dielectric constant will have
a periodic spatial variation whose wavelength is equal to the phonon
wavelength. This acts as a diffraction grating for light, moving
moreover with a velocity $v_q = \omega_q \lambda_q$, where the subscript q refers
to the direction vector of the phonons. The frequency ω_s of the
scattered light will therefore have suffered a Doppler shift given by

$$\omega_i - \omega_s = \pm \omega_q$$

Thus the scattered light is shifted to higher and lower frequencies
by an amount corresponding to the phonon frequency. Alternatively, Brillouin scattering can be viewed quantum mechanically
as an inelastic scattering event between a photon and a phonon,
the kinematics of the collision process leading to the same result.

Thus the direction of the incident light beam and the direction
of viewing of the scattered light select two waves of frequency ω_q,
travelling in opposite directions, from all the acoustic waves
present, the frequency of the phonons investigated increasing
with scattering angle. The maximum phonon frequency, probed
by back scattering of the light, falls in the microwave range, so that
Brillouin scattering would seem eminently suitable for probing
the phonons in a microwave phonon bottleneck, and the proposed
experiment is as follows.

Light is passed into the crystal from a laser, and that scattered
through 90° is observed using a Fabry–Perot interferometer, the
pass-frequency of which may be swept by varying the separation
between the plates, using piezoelectric transducers mounted on

one plate to move it relative to the fixed plate at some low audio frequency. This enables use to be made of a signal averaging technique and allows the frequency spectrum of the Brillouin scattered light to be scanned.

The laser light is reflected back through the specimen by means of a mirror at the far end (see Fig. 8.16) so that phonons labelled q_A, which are observed from the laser light travelling to the left, as well as those q_B, observed from the laser light travelling to the right, may be investigated.

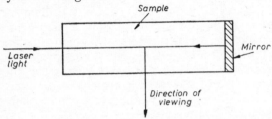

FIG. 8.16. Experimental arrangement for viewing Brillouin scattered light from a sample (Geschwind[18]).

The spin system is saturated with microwave radiation at the spin resonance frequency (in the applied magnetic field). These spins relax, and heat up the phonons in a narrow band at this frequency. The intensity of q_A will therefore increase, while q_B remains unchanged. q_B is therefore used to monitor the temperature rise of the crystal, if any, as opposed to the selective heating of q_A by the relaxing spins. It is anticipated that a rise in temperature of 0·1°K should be detectable in selected phonons in MgO, whereas bottlenecks are known to heat up the phonons by several degrees (see p. 72). By altering the direction of viewing, it should be possible to examine the relative heating of all the different phonons at that resonant frequency.

This ability to select phonons of a given frequency and **k**-vector, and to examine how the relaxing spins feed these different phonons is what renders unique the examination of the microwave phonon bottleneck by Brillouin scattering.

In conclusion, it would seem that optical techniques are opening up new fields for investigating spin-phonon interactions, both in excited and ground states.

§8.4

DETERMINATION OF T_1 FROM E.S.R. LINEWIDTH MEASUREMENTS

By the uncertainty principle, the existence of a spin-lattice relaxation time T_1 must imply a linewidth of order $1/T_1$. This will not be the only line broadening mechanism and it is not always feasible to isolate the contribution of $1/T_1$ to the observed linewidth. However, if we have a system with a limiting linewidth ΔH_0 at low temperatures and the line shape is Lorentzian, there is then good reason to suppose that the broadening is homogeneous. If the line broadens further as the temperature is raised, but accurately retains its Lorentzian shape, it is probable that homogeneous broadening is still occurring and that the $1/T_1$ contribution has been added linearly. Thus

$$\Delta H_m = \Delta H_0 + \Delta H_{T_1}$$

where ΔH_m is the measured linewidth, the overall width at half power. ΔH_{T_1} is the contribution at that temperature from spin-lattice relaxation and T_1 may be calculated from

$$T_1 = h/\pi g \beta \Delta H_{T_1}$$

Assuming $g = 2$, a relaxation time of 1 μsec gives a linewidth contribution of about 0·1 gauss, while $T_1 = 10^{-9}$ sec corresponds to $\Delta H_{T_1} \approx 110$ gauss.

This method is unlikely to be reliable unless (*a*) the limiting value of the linewidth ΔH_0 can be measured, (*b*) ΔH_0 does not arise from temperature dependent causes, (*c*) the line shapes are accurately Lorentzian, and (*d*) T_1 is less than about 10^{-8} sec. It is, however, often the only method available for estimating those short relaxation times which frequently occur at the higher temperatures where the Raman mechanism dominates the relaxation process.

References

1. Hahn, E. L., *Phys. Rev.*, 1950, **80**, 580.
2. Portis, A. M., *Phys. Rev.*, 1953, **91**, 1071.
3. Mims, W. B., *Rev. Sci. Inst.*, 1965, **36**, 1472.
4. Klauder, J. R., and Anderson, P. W., *Phys. Rev.*, 1962, **125**, 912.
5. Mims, W. B., Nassau, K., and McGee, J. D., *Phys. Rev.*, 1961, **123**, 2059.
6. Dyment, J. C., *Can. Jour. Phys.*, 1966, **44**, 637.
7. Breen, D. P., D.Phil. thesis, Oxford University, 1966.
.8 Bloch, F., *Phys. Rev.*, 1946, **70**, 460.
9. Abragam, A., *The Principles of Nuclear Magnetism* (O.U.P., 1961).
10. Collins, S. A., Kyhl, R. L., and Strandberg, M. W. P., *Phys. Rev. Letts.*, 1959, **2**, 88.
11. Castle, J. G., Chester, P. F. and Wagner, P. E., *Phys. Rev.*, 1960, **119**, 953.
12. Rannestad, A., and Wagner, P. E., *Phys. Rev.*, 1963, **131**, 1953.
13. Brya, W. J., and Wagner, P. E., *Phys. Rev.*, 1967, **157**, 400.
14. Firth, I. M., and Bijl, D., *Phys. Letts.*, 1962, **2**, 3, 160.
15. Geschwind, S., Devlin, G. E., Cohen, R. L., and Chinn, S. R., *Phys. Rev.*, 1965, **137**, A1087.
16. Addé, R., Geschwind, S., and Walker, L. R., *Proceedings of the XV Colloque Ampère*, 1968.
17. Imbusch, G. F., and Geschwind, S., *Phys. Rev. Letters*, 1966, **17**, 238.
18. Geschwind, S., *Proceedings of the XV Colloque Ampère*, 1968.

Chapter 9

Experimental Spin-Lattice Relaxation Time Data

In the foregoing chapters the theory of spin-lattice relaxation has been outlined and some experimental methods for determining the relaxation parameters have been discussed. Ten years have passed since the first pulse relaxation measurements and it is our purpose in this chapter and in Appendix 2 to collect together and comment upon some of the results which have been published in that time. The relaxation of the Cr^{3+} ion in Al_2O_3 (ruby) has been the subject of extensive investigation, and in §9.1 some of the more important facets of the results are discussed. Because of the number and thoroughness of the investigations, ruby is a very appropriate example to illustrate the behaviour of transition metal ions. The next section, §9.2, is concerned with a detailed comparison of experiment and theory for a representative selection of rare earth ions. In Appendix 2, other published results are given, with a few details of the experimental conditions and with such comment as is appropriate.

§9.1

THE Cr^{3+} ION IN Al_2O_3—RUBY

In 1958 Makhov[1] *et al.* reported the successful observation of maser action in ruby, and there followed a series of papers on this topic with ruby as the active material. At this time it seemed that the maser might make an important contribution to communications by satellite. Since relaxation plays an important role in determining the efficiency of maser operation, the spin-lattice relaxation parameters in ruby were carefully studied by a number of workers.

One of the first comprehensive papers was that of Gill.[2] He measured relaxation rates at 9·5 Gc/s using a pulse technique, and his ruby samples had a nominal Cr–Al ratio extending from 0·01 to 0·76 per cent. The crystals were grown by a flame-fusion method.

TABLE 9.1. T_1 measurements on ruby. $T = 4\cdot2°$K.

Author and reference	Frequency (Gc/s)	Saturating pulse width (msec)	Transition (see Fig. 9.1)	$\theta°$	Cr^{3+} concentration (atomic %)	T_1 (msec)	Remarks
Davis et al.[17]	8·75	0·001	1-2	90	0·03	37	
Armstrong and Szabo[18]	2·92	0·2	1-2	0--90	0·03	300	T_1 varied with θ from 100 msec to 1 sec
Mims and McGee[19]	7·17	0·50-1·0	1-2	90	0·02 0·15	500 100	Cross-relaxation study
Pace et al.[3]	34·6	0·05-5·0	1-2	90	0·03	22	Other transitions measured
Gill[2]	9·5	0·1 or greater	1-3	0	0·01 0·03 0·08	300 150 65	Relaxation times for 'pairs' also measured
Manenkov and Prokhorov[5]	9·4	0·8-1000	1-3	0	0·05 0·10 0·40 0·65	98 64 20 10	Other transition measured
Nisida[4]	9·3	0·1-1·0	1-3	0	0·023 0·14	220 50	Concentration range 0·009-0·79% studied
Feng and Bloembergen[20]	8·4	'long'	1-3	0	0·04	100	D.C. magnetization method, other concentrations and transitions studied
Standley and Vaughan[8]	9·27	up to 200	1-3	0	0·005 0·20	150 105	See text and Table 9.2
Mason and Thorp[12]	35	up to 100	2-3	90	0·013 0·052	155 175	Concentration range 0·002-0·2% studied
Lees et al[9]	35	up to 100	3-4 2-3	0 0	0·017 0·20 0·017 0·20	64 36 39 16	D.C. magnetization method

He measured the relaxation both of single ions and of pairs, and it is with the former measurements that we are primarily concerned, although reference will be made to the pair measurements below. The recovery in the more concentrated crystals was found to be modified by cross relaxation, as would be expected. Indeed, attempts were made to measure the cross-relaxation times and they varied from less than 5 μsec for the most concentrated specimen to 150 μsec in a sample containing about 0·2 per cent Cr^{3+}. Because of the presence of general cross relaxation, the time constant measured from the last (detectable) part of the recovery had to be taken as a measure of the spin-lattice relaxation time. The duration of the saturating pulse was increased to aid measurement of components of small amplitude and long time constant. The measurements were taken in the temperature range 1·7 to 100°K and evidence was found of both the direct and Raman processes. As shown in Table 9.1, a most significant feature of Gill's results was a clear dependence of relaxation time on concentration. Furthermore, the relaxation times were generally about an order of magnitude greater than those predicted by theoretical analyses of the type given in Chapter 2.

Pace *et al.*[3] reported measurements at 34·6 Gc/s also using a pulse-saturation technique. Some of these results are given in the table and show again a variation of relaxation time with concentration. In all cases T_1 was of the order 100 msec at 4·2°K. The authors found a T^{-1} temperature dependence of T_1 between 1·4 and 20°K, with a more rapid decrease in relaxation time above this temperature, consistent with the onset of the Raman mechanism. At 1·4°K, the values of T_1 were markedly dependent on the concentration of chromium in the material but the variation was within experimental error at 77°K.

Nisida[4] worked with chromium ion concentrations ranging from 0·001 to 0·8 atomic per cent. The method of preparation of the samples is not stated but they were probably formed by the flame-fusion technique. The results obtained were similar to those of Gill in the same frequency region. Cross-relaxation effects were observed and attempts were made to reduce their influence on the T_1 measurements. Values of T_1 were found to be independent of concentration for samples containing less than 0·01 per cent Cr^{3+}, but in other samples a concentration dependence was found reaching approximately (concentration)$^{-2}$ in rubies containing most chromium.

Manenkov and Prokhorov[5] found that for a sample containing 0·15 per cent Cr^{3+} at 4·2°K, T_1 varied from about 35 to 96 msec, depending on crystal orientation, while the cross-relaxation time (chapter 3) varied from 2 to 4 msec. These results were obtained using a long saturating pulse of order 460 msec. With saturating pulses 0·6 msec long, values of T_1 were the same order of magnitude although some changes were observed for particular transitions and angles. These authors also found that the relaxation time was a function of chromium concentration (see Table 9.1).

In 1964, Donoho[6] published calculations of the relaxation times in ruby at 9·3 Gc/s. The method of calculation was not unlike that outlined in Chapter 2 and was based on a spin-lattice inter-action Hamiltonian due to Van Vleck, namely

$$\mathcal{H}_{SL} = \sum_{ij} D_{ij} S_i S_j$$

The tensor D is defined as

$$D_{ij} = \sum_{ijkl} G_{ijkl} e_{kl}$$

and e_{kl} is a lattice strain given by

$$e_{kl} = \frac{1}{2} \left[\frac{\delta u_k}{\delta x_l} + \frac{\delta u_l}{\delta x_k} \right]$$

where u is the displacement of the atoms in the unit cell caused by the lattice waves. These equations define G as a coefficient which, multiplying the strain, gives the change in crystal-field energy under lattice strain. Values of the components of the G tensor obtained by Hemphill and Donoho[7] from static strain measurements were used to compute the phonon-induced transition probabilities for the direct relaxation process between the various spin levels in ruby. These transition probabilities W_{ij} were then used to solve the rate equations for the four-level ruby spin system, the equations being of the form

$$\frac{dn_i}{dt} = \sum_{j=1}^{4} (W_{ji} n_j - W_{ij} n_i) \quad i = 1, 2, 3, 4$$

where n_i is the population of level i. Such a set of equations in general yield three independent relaxation parameters for a four-level system, and a solution is obtained of the general form

$$\frac{n}{n_0} = 1 + A_1 \exp(-t/\tau_1) + A_2 \exp(-t/\tau_2) + A_3 \exp(-t/\tau_3)$$

Here n is the instantaneous value of the population difference of the two levels between which resonance is being observed, and n_0 is its thermal equilibrium value. The parameters τ_1, τ_2, τ_3 are three time constants, not simply related to the $T_1(ij)$ defined for the three pairs of levels i and j, and A_1, A_2 and A_3 are the relative amplitudes of the contributions from these terms.

Values of A_1, A_2 and A_3 and of τ_1, τ_2, and τ_3 at $4 \cdot 2°$K were computed numerically by Donoho and presented as a function of the angle between the applied magnetic field and the crystal

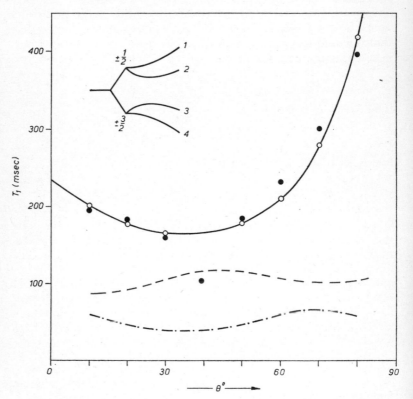

Fig. 9.1. Comparison of experimental and calculated relaxation times for the (2-3) transition in ruby.[8] The solid curve and the ○'s represent the calculated values and the ●'s are experimental points for VP rubies. The two broken curves refer to V samples. The Donoho nomenclature for the ruby energy levels is given in the inset.

c-axis for the four transitions; his notation for the levels is shown in Fig. 9.1. There was a strong angular dependence of the relaxation times predicted by Donoho; this is a consequence of the variation in admixture between the states as the angle θ is changed. In most cases he found that one relaxation time, usually the longest, dominated the behaviour; when two relaxation times were important, they were usually nearly equal. It was found that A_3, the amplitude of τ_3, was usually negligible compared with A_2 and A_1.

Reference is made in Chapter 7, p.124, to the work of Standley and Vaughan[8] in comparing the experimental results with Donoho's predictions. Fig. 7.10 shows the curves of predicted relaxation times against angle for the (3–4) transition, while Fig. 7.11 shows relative values of A_1 and A_2 for three transitions. The ruby samples, obtained from a number of sources, had been made by three techniques;

(i) *The Verneuil process*, in which solid oxides of aluminium and chromium are fed to an oxy-hydrogen flame and the resulting molten mixture crystallizes in a boule on a seed crystal below the flame tip.

(ii) *The so-called 'vapour phase' modification of the Verneuil process*, developed by Thermal Syndicate Ltd, wherein aluminium and chromium are fed to the flame as halide vapours instead of as solid oxides.

(iii) *The flux-melt technique*, in which the ruby crystals grow from a solution of the oxides in lead oxide or lead fluoride flux. This process permits lower temperature growth of the rubies.

Below we refer to these three techniques as V, VP and FM respectively.

The (2–3) high-field transition is particularly relevant because Donoho predicted a very strong angular dependence of T_1 as shown in Fig. 9.1. A_2 and A_3 were found to be negligible compared with A_1 for this transition, and only one time constant is expected, namely, τ_1 which one therefore identifies with T_1. This was found to be the case experimentally at 9·3 Gc/s, so long as one was working with VP rubies. The results obtained using this material are shown in Fig. 9.1 and they are in excellent agreement with the Donoho predictions with regard to the form of the angular dependence, but in order to obtain good *numerical* agreement, it was necessary to reduce the magnitudes of the predicted values of T_1 by a factor of 2; in view of the approximations used in the calculations, the agreement is still good. It will be noted that there is a single experimental point at 40° which does not fit the theoretical curve at all well; this may be explained in terms of harmonic cross relaxation, for $\nu_{23} = 2\nu_{34}$ at about 42°. A harmonic point at about 83° ($\nu_{23} = \nu_{34}$) may account for the experimentally-obtained value of T_1 at 80° being apparently slightly low, although this is within the experimental error. All the recoveries for this transition, except close to the harmonic angles, were exponential in form and showed no dependence on pulse width when this was varied between 15 μsec and 200 msec. The angular dependence of T_1 at 1·6°K also followed a similar form with the value of T_1 increased according to $T_1^{-1} = AT$. Data similar to these were obtained with all VP crystals containing not more than 0·2 per cent Cr^{3+}.

This type of behaviour was not found with crystals produced by the V or FM processes. With these, very little variation in relaxation time with crystal orientation was observed, in agreement with the findings of Gill. This is strikingly shown in Fig. 9.1 where the results are given for two V crystals containing 0·05 and 0·2 per cent Cr^{3+}.

TABLE 9.2. Results for some VP rubies measured at 9·27 Gc/s

Concentration at. % Cr^{3+}	Relaxation time T_1		
	1·6°K (msec)	4·2°K (msec)	77°K (μsec)
0· 005	370	150	
0·017	540	230	27
0·018	500	190	15–20
0·045	550	240	
0·047	350	140	33
0·050	540	200	
0·087	550	215	29
0·20	420	105	30

As Table 9.2 shows, with the VP crystals there was no clear dependence of T_1 on concentration either at 1·6 or 4·2°K for crystals containing not more than 0·2 per cent Cr^{3+}. V and FM crystals showed a concentration dependence of the type reported by other workers.

The temperature dependence of T_1 for four VP rubies is shown in Fig. 9.2, where the full line is drawn for $T_1^{-1} = AT$. This behaviour was found by other workers when concentration independent values were obtained. There is no *a priori* reason why rubies having the same chromium concentration but made by different techniques should have different relaxation times. Thus if we assume (and the agreement with Donoho's predictions makes this reasonable) that in the VP rubies one is truly measuring T_1 for the Cr^{3+} ion, then in the V rubies at helium temperatures one is not observing the *direct* relaxation of that ion to the lattice. Lees, Moore and Standley[9] used the same rubies as Vaughan but worked at 35 Gc/s. There was now far less difference in the values of T_1 for the three types of sample, and there was no clear dependence of T_1 on orientation, though this is still predicted by Donoho at the higher frequency. Fig. 3.9 on p. 54 suggests one reason. The very high possibility of harmonic cross relaxation makes comparison between the experiment and a theory which excludes cross relaxation quite unprofitable.

In the theoretical discussion of Chapter 2 the magnetic ion was considered to be completely isolated and therefore no question of

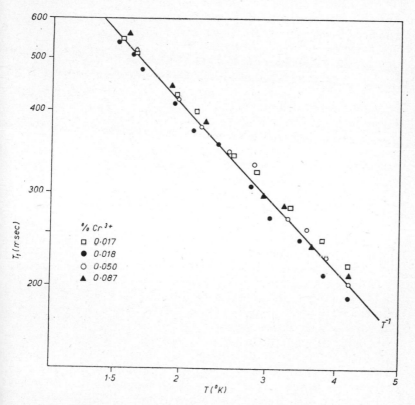

FIG. 9.2. The temperature dependence of T_1 for four VP rubies.[8]

concentration dependence can come into the calculations. One would expect no concentration dependence to be found experimentally so long as the amount of Cr^{3+} present were such that the theoretical conditions were approximately true. With 0·1 per cent added, one would expect on average to move some ten atomic spacings from one magnetic ion before meeting another, and one might reasonably suppose that such a magnetic ion were well

N

isolated. The experiments with Verneuil rubies show that concentration dependence is found at a much lower concentration
than this, whereas with the VP rubies the concentration dependence sets in at about 0·2 per cent. It is interesting to speculate on
the cause of these phenomena. The most likely solution is that the
Cr^{3+} ion is cross-relaxing to another magnetic impurity which is
itself tightly coupled to the lattice to which it relaxes very rapidly.
The relaxation of the Cr^{3+} ion is monitored, without there being
a simple test of whether it is wholly a true spin-lattice relaxation
which is being observed. The spectrum of the fast relaxer must
obviously be very broad because it must overlap with the Cr^{3+}
line over a wide variation of angle. This type of postulate satisfactorily explains both the shorter relaxation time and the lack of
variation with angle which one sees with the V rubies.

At the time of writing the nature of the fast relaxer is not certain.
It has been suggested that the observed concentration-dependence
occurs when relaxation takes place via exchange-coupled pairs or
groups of ions. In drawing attention to this possibility, Van Vleck[10]
pointed out that the spectrum produced by pairs is quite complex
and he envisaged the relaxation mechanism as a modified Orbach
two-phonon process taking place via excited levels in the pair
spectrum. Gill[2] calculated the effect of relaxation via pairs of ions
and concluded that below a concentration of about 0·2 per cent
Cr^{3+}, the coupling is too weak and the pair lines too few to account
for the observed behaviour. The effect may, however, in his view
influence the relaxation behaviour of more concentrated specimens.
It is not clear that proper account has necessarily been taken of
higher order groups, namely three or more ions, nor of possible
effects due to non-uniform doping of the Al_2O_3 by the Cr^{3+}.
Microprobe analysis has shown that the chromium is not uniformly
distributed through the lattice and there may be sufficient 'clumps'
of chromium in the material to provide the fast relaxing centres.
So far as the effect of pairs is concerned, Saunders, Standley and
Wilson[11] measured the pair spectra of the samples used by Standley
and Vaughan in the relaxation work and three specimens of similar
chromium concentration were found to have pair spectra which
were very similar both in form and intensity. One of these was a
VP ruby with a marked angular dependence of T_1, while the others
were V and FM rubies with no clear dependence on angle and with
a measured value of T_1 three times smaller than that of the VP
specimen. The occurrence of pairs of ions is therefore unlikely to
be the main cause of the short relaxation times in this case.

Mason and Thorp[12] examined the effect of crystalline imperfections on T_1 in ruby. The relaxation times were measured at

35 Gc/s using a pulse-saturation technique, while crystallographic imperfections were studied by etching and X-ray methods. The dislocation densities were determined by counting etch pits produced on (0001) faces after etching the specimens in orthophosphoric acid at 320°C. Two parameters were determined by X-ray techniques—the first was mosaic misorientation revealed by the occurrence of several closely grouped spots instead of a single

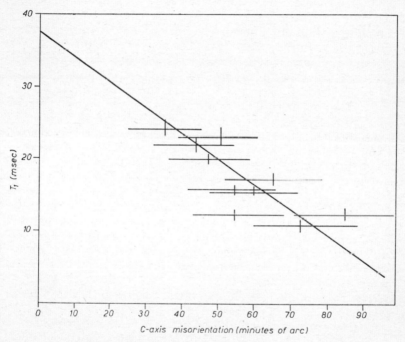

FIG. 9.3. The dependence of T_1 (measured at 35 Gc/s) upon c-axis misorientation in ruby (after Mason and Thorp[12]).

diffraction spot on a film, while the second, the c-axis misorientation, represented the extent of the variation in the mean direction of the crystallographic c-axis over a specimen and was estimated by comparison of photographs taken at different parts of the specimen. Fig. 9.3 shows the measured relaxation time plotted against c-axis misorientation for a number of rubies. Mason and Thorp concluded that crystalline imperfections may affect the magnitude of the relaxation times by at least a factor of 2 in typical specimens; such a variation does not seem to depend on method of manufacture.

It has been suggested that the fast relaxing impurity may be a magnetic ion which should not be present. That is to say, it could be a true impurity ion of a type other than chromium, or it might be chromium in a valence state other than the trivalent one. Ferric iron, Fe^{3+}, is very often present in ruby crystals. The relaxation time of this ion has been measured, however, and it is found to be the same order as that of the Cr^{3+} ion. Moreover, its lines are sharp and there is no possibility of overlap at all angles. Another possibility is Fe^{2+} which is known to enter the lattice in the growth of sapphire if one attempts to increase the Fe^{3+} content above a certain level. In the work of Standley and Vaughan, the total iron content was measured spectroscopically and there was as much iron present in the V rubies as in the VP ones. It would be surprising if some of this were not in the Fe^{2+} state and therefore it is felt that Fe^{2+} is not the most likely candidate for the fast relaxer. This has now been confirmed by Stevens and Walsh[13] from theoretical considerations. We are left therefore with chromium. Hoskins and Soffer[14] reported the spectrum of Cr^{4+} in the Al_2O_3 lattice, a single line with $g_{\parallel} = 1·9$, certainly not broad enough to overlap all the Cr^{3+} transitions. Using an orange ruby known to contain both Cr^{3+} and Cr^{4+} Standley and Vaughan[15] measured the values of T_1 for Cr^{3+} throughout a region in which the Cr^{4+} resonance crossed that of the Cr^{3+}. The conditions for cross relaxation were thus very good but little change in T_1 could be found. It is less easy to be certain about Cr^{2+} although the evidence from microwave phonon work is that this again is not likely to prove to be the fast relaxing impurity.

The position then is still somewhat obscure, for to date the fast relaxer has not been positively identified. Its presence is the only satisfactory explanation of the difference in behaviour of the different types of ruby and no doubt further experimental work will reveal its nature in due course. Some of the published results on ruby are summarized in Table 9.1. Excluded from the table are early results employing the continuous saturation method, for these are particularly difficult to assess. One may comment that using this technique, Manenkov and Prokhorov[16] found $T_1 = 44$ msec at a frequency of 9·4 Gc/s for $\theta = 0$ at $4·2°K$, the material having a chromium concentration of 0·05 atomic per cent. This is clearly the same order of magnitude as those in the table and the authors also reported a concentration dependence.

The work on ruby has been important, for it has clearly shown the need for publication of a full description of the materials and of the experimental techniques and conditions if one is to get reliable T_1 data from the literature in future.

§9.2

SOME RARE-EARTH IONS

In 1961 Orbach[21] published his phenomenological theory of spin-lattice relaxation. This was most applicable to rare-earth ions, and much experimental work has since been devoted to the study of these ions in a variety of host lattices. It has been found that in most cases their behaviour follows very closely that predicted by Orbach's theory; unlike ions of the iron group, disagreement is the exception rather than the rule. There are several reasons why this might be so:

(i) J is a good quantum number, so that fewer approximations need be made in the theory.

(ii) Linewidths are generally small compared with the fields at which measurements are made.

(iii) Resonance lines are usually well separated so that cross-relaxation effects are minimal.

(iv) Probably most important, the orbit-lattice interaction is so strong that impurities do not generally cause trouble.

Crystallographic data and crystal-field parameters are often quite accurately known for these ions, since much optical work has been done on them, and the energy-level splittings are often such that the first excited level occurs at an energy less than the Debye cut-off energy, $k\theta_D$. This means that phonons are present in the lattice which are able to participate in the resonant two-phonon process, the Orbach process, and indeed this mechanism has often been observed. Several rare-earth ions also exhibit a phonon bottleneck in certain host lattices, and a study of these materials can afford a very convincing confirmation of the validity of the theory of Faughnan and Strandberg,[22] and of Scott and Jeffries.[23]

In order to illustrate some of the above points, two specific examples have been chosen from the many reports of such measurements. In Appendix 2 an attempt has been made to summarize most of the other published results.

Nd^{3+} in the double nitrate $La_2Mg_3(NO_3)_{12}.24H_2O$

A very detailed study of the relaxation behaviour of trivalent neodymium in the lanthanum magnesium double nitrate (LaMN) was reported in 1962. Scott and Jeffries[23] measured T_1 in the temperature range $1.4 \to 4.2°K$ at 9.27 Gc/s and at 34 Gc/s, while Ruby *et al.*[24] extended the temperature range down to $0.3°K$ at 9.67 Gc/s. Fig. 9.4 shows the experimental results of both investigations (at the lower frequency), obtained for the strong central line ($I = 0$) of the Nd^{3+} spectrum, with $H \perp z$-axis, for two different LaMN crystals, the first containing 5 per cent Nd^{3+} of natural abundance, and the second containing 1 per cent Nd^{3+}

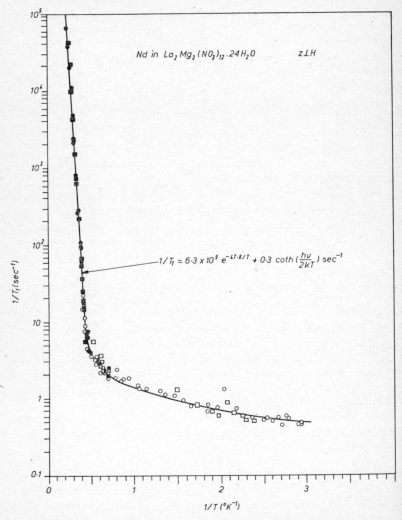

Nd in $La_2 Mg_3 (NO_3)_{12} . 24H_2O$ $z \perp H$

$1/T_1 = 6.3 \times 10^9\, e^{-47.6/T} + 0.3\, \coth\left(\frac{h\nu}{2kT}\right)\, \sec^{-1}$

FIG. 9.4 Relaxation data for Nd^{3+} in LaMN (Ruby *et al.*[24]).

⊛ 5% Nd ⎫
■ 1% Nd ⎭ Conventional apparatus

○ 5% Nd ⎫
□ 1% Nd ⎭ Demagnetization apparatus

enriched to 98·5 per cent of the even isotopes. For both crystals, the curve which best represents the data is given by

$$\frac{1}{T_1} = 6\cdot3 \times 10^9 \exp(-47\cdot6/T) + 0\cdot3 \coth(h\nu/2kT) \sec^{-1}$$

At the time of its publication, this was the first set of data to require the term $A \coth(h\nu/2kT)$; usually $h\nu$ is so much smaller than kT that this term approximates closely to AT (see p. 12). Indeed, Scott and Jeffries found that the data above 1·4°K could be represented by a term 1·7 T.

An Orbach two-phonon relaxation process is clearly indicated. The value of 47·6°K in the exponent was chosen to agree with the optically measured field splitting for NdMN, but the relaxation data were still sufficiently precise to verify this value to ±1°K for these dilute crystals. No concentration dependence was observed, and cross relaxation between hyperfine lines appeared to be negligible.

As T becomes small compared with $h\nu/k$, it is seen from Fig. 9.4 that the direct relaxation rate tends towards the temperature independent value $1/T_1(d) = K = 0\cdot3 \sec^{-1}$, which is just the transition probability for spontaneous emission of phonons. This can easily be shown from the arguments of Chapter 1. From equation (1.13), p. 11, as $T \to 0$, then $P(\delta) \to 0$, and only the second term is left in equation (1.17). Hence at absolute zero, under the experimental conditions reported by Ruby *et al.* [24], the direct relaxation time $T_1(d) = K^{-1} \approx 3\cdot3$ sec.

Scott and Jeffries note that several runs on a given crystal always gave very reproducible results, but that two different crystals (even from the same batch of growing solution) gave values of T_1 often differing by as much as a factor of two. The discrepancy between the above results and a value of $2\cdot1 \times 10^9 \exp(-47\cdot6/T)$ obtained for a concentrated crystal was not considered significant. Cowan *et al.* [25] obtained a value $1/T_1(0) \approx 2 \times 10^9 \exp[(-46 \pm 2)/T] \sec^{-1}$ by a spin-echo method for a crystal of 0·2 per cent Nd^{3+} in LaMN.

Scott and Jeffries calculated values for $1/T_1(d)$ and $1/T_1(0)$ in a manner very similar to the example shown in Chapter 2. The wave functions they used, particularly for the excited doublet, were very approximate, as were their values for the velocity of sound in the crystal. Their predicted expression for the relaxation rate

$$\frac{1}{T_1} \approx 2\cdot6T + 2\cdot2 \times 10^{10} \exp(-47\cdot6/T) + 7\cdot8 \times 10^{-4}T^9 \sec^{-1}$$

is in remarkably good agreement with their experimental value.

The Raman term is at least an order of magnitude smaller than the others, and hence would not be detected.

No phonon bottleneck was observed at X band even at the lowest temperature. Measurements made on the 1 per cent Nd crystal at 34·3 Gc/s, however, fitted the expression

$$T_1^{-1} = 32T^2 + 4 \times 10^9 \exp(-47·6/T) \sec^{-1}$$

The T^2 dependence of the first term is characteristic of a phonon bottleneck and Scott and Jeffries showed that these results were in close accord with bottleneck theory (see Chapter 4, p. 72, for an outline of their arguments).

In an experiment to check the effect of paramagnetic impurities on relaxation behaviour, Scott and Jeffries made measurements on the Nd line at 9·37 Gc/s in a LaMN crystal containing 1·5 per cent Ge and 0·2 per cent Pr. The data fitted the expression

$$\frac{1}{T_1} = 6·2T + 7·6 \times 10^9 \exp(-47·6/T) \sec^{-1}$$

showing that the direct process, at least, is sensitive to small amounts of other paramagnetic impurities, particularly if they have a faster relaxation rate (in the case of the Pr ion the rate is faster by $\sim 10^5$). Further confirmation is found in the results for Nd in LaES reported in the same paper.

Ce^{3+} in the double nitrate LaMN

Fig. 9.5 shows the results of several different workers for Ce^{3+} in LaMN in the temperature range 0·25 → 4·2°K. Above about 1·4°K there is excellent agreement between the different sets of results, obtained using several different methods, and an Orbach process is predominant. In this region there is no dependence on concentration.

Ruby *et al.* extended the measurements down to about 0·25°K, using an adiabatic demagnetization technique for cooling, where they found their results to be concentration dependent, as shown in the figure. The continuous lines in this figure represent

$$\frac{1}{T_1} = 2·7 \times 10^9 \exp(-34/T) + D'\coth^2(h\nu/2kT) \sec^{-1}$$

where $D' = 0·8$ for the 0·2 per cent Ce^{3+} specimen and 2·4 for the 2 per cent specimen. This $\coth^2(h\nu/2kT)$ type of dependence is expected in the case of a phonon bottleneck when $h\nu \ll kT$ (see p. 64).

Using reasonable values for the wavefunctions and other parameters, Ruby *et al.* calculated for the Orbach term

$$\frac{1}{T_1(0)} \approx 3·5 \times 10^9 \exp(-34/T) \sec^{-1}$$

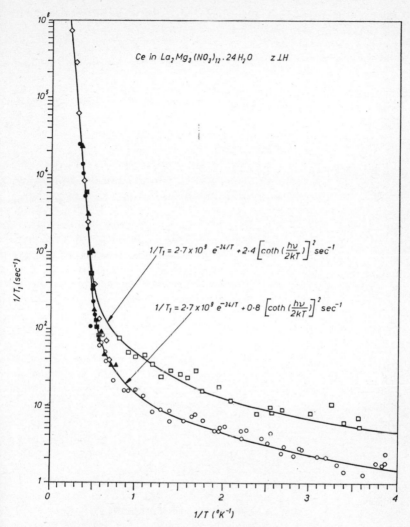

Ce in $La_2Mg_3(NO_3)_{12}.24H_2O$ $z \perp H$

$$1/T_1 = 2.7 \times 10^9 \, e^{-34/T} + 2.4 \left[\coth \left(\frac{h\nu}{2kT} \right) \right]^2 \text{sec}^{-1}$$

$$1/T_1 = 2.7 \times 10^9 \, e^{-34/T} + 0.8 \left[\coth \left(\frac{h\nu}{2kT} \right) \right]^2 \text{sec}^{-1}$$

Fig. 9.5 Relaxation data for Ce^{3+} in LaMN (Ruby *et al.*[24]).

● 100 per cent Ce, Finn *et al.*[26] (non-resonant method).

■ 5 per cent Ce, Leifson and Jeffries[27] (pulse-saturation method).

△ 1 per cent Ce, Scott (private communication referred to in reference 24) (pulse-saturation method).

◇ 0.2 per cent Ce, Cowen and Kaplan[28] (spin-echo method).

○ 2 per cent Ce
□ 0.2 per cent Ce $\Big\}$ Ruby *et al.*[24] (pulse-saturation method)

which is in very good agreement with the experimental value. For the direct term, however, they calculated

$$\frac{1}{T_1(\mathrm{d})} \approx 1 \cdot 3 \coth(h v/2kT) \sec^{-1}$$

in contrast with the lower limit of this term which was estimated from the experimental data as

$$\frac{1}{T_1(\mathrm{d})} = 20 \coth(h v/kT)$$

This is surprising, because even though order of magnitude discrepancy between calculated and measured values is sometimes found for other ions, it is usual to find that the experimental *ratios* for two different processes are correctly predicted by the theory. Possible explanations for this discrepancy could be that the wave functions used were not sufficiently accurate, and that account should have been taken of the presence of higher energy levels. Nevertheless, the results for this salt do show that it is possible to obtain good agreement between measurements made by different methods and using different samples.

Critique of the theory

Huang[29] has criticized Orbach's phenomenological orbit-lattice interaction on two grounds. Firstly, the approximation that the dynamic field interaction is approximately equal to the static, experimentally known, value (see p. 22) is a poor one in that, if a point-charge model is used, A_2^m is proportional to ξ^{-3}, A_4^m to ξ^{-5} and A_6^m to ξ^{-6}. Hence

$$\left| \xi \left(\frac{\delta A_n^m}{\delta \xi} \right)_0 \right| = q_n |A_n^m(\xi_0)|$$

where $q_n = 3, 5, 7$ for $n = 2, 4, 6$ respectively. The error involved in taking $q_n = 1$ (as we did in Chapter 2) may be very serious. For example, when $n = 6$ terms are dominant, the relaxation rate calculated for a two-phonon Raman process could be in error by three orders of magnitude.

Secondly, Huang points out that, by Orbach's method, it is not possible to estimate a value for the parameter $A_n^m r^n$ when $n = 2$ for the paramagnetic ion situated on a cubic site. He therefore concludes that Orbach's theory is easier to use for crystals of lower symmetries for which the normal co-ordinates are harder to find, and that for a crystal of cubic symmetry one should use the Van Vleck theory.

TABLE 9.3. Comparison of measured data and theoretical estimates of the spin-lattice relaxation rates for rare-earth ions in single crystals.[29]

Salt		Direct process T_1^{-1} (sec^{-1})	Orbach process T_1^{-1} (sec^{-1})	Raman process T_1^{-1} (sec^{-1})
0·2% Tm^{2+} in CaF$_2$	Expt.	13T		7·7 × 10$^{-8}T^9$
	Theor.	4·5T		1·9 × 10$^{-6}T^9$
0·02% Ho^{2+} in CaF$_2$	Expt.	42T	8·0 × 10^9 exp(−48/T)	
	Theor.	37T	8·5 × 10^9 exp(−48·7/T)	
0·0047, 0·18, 0·0074T Eu^{2+} in CaF$_2$	Expt.	12T		5·3 × 10$^{-4}T^5$
0·1% Yb^{3+} in YGaG	Expt.	33T		1·8 × 10$^{-7}T^9$
	Theor.	4·2T		5·7 × 10$^{-7}T^9$
0·1% Yb^{3+} in YAlG	Expt.	15T		6·3 × 10$^{-7}T^9$
	Theor.	5·3T		9·0 × 10$^{-7}T^9$
1% Nd^{3+} in YGaG	Expt.	17T	9·0 × 10^{10} exp(−120/T)	
	Theor.	10T	10^{12} exp(−120/T)	
1% Nd^{3+} in YAlG	Expt.	34T	4·5 × 10^{10} exp(−110/T)	
	Theor.	10T	10^{11} exp(−110/T)	
0·1% Sm^{3+} in LES	Expt. $H_\parallel$	16T		2·6 × 10$^{-2}T^9$
	Theor. $H_\parallel$	0·32T		4·4 × 10$^{-3}T^9$
	Expt. $H_\perp$	100T		2·6 × 10$^{-2}T^9$
	Theor. $H_\perp$	1·8T		4·4 × 10$^{-3}T^9$
	Expt. $H_{45°}$	19T		2·6 × 10$^{-2}T^9$
1% Yb^{3+} in YGaG	Expt.	30$T^{1·4}$		1·8 × 10$^{-7}T^9$
10% Yb^{3+} in YGaG	Expt.	410T		1·8 × 10$^{-7}T^9$
1% Yb^{3+} in LGaG	Expt.	9·8$T^{1·7}$		1·0 × 10$^{-7}T^9$
1% Yb^{3+} in YAlG	Expt.	11$T^{2·3}$		6·3 × 10$^{-7}T^9$

Huang studied some rare-earth ions in the lattice CaF_2 and also in some garnets, all of which have XY_8 symmetry. He used Orbach's symbolism, but Van Vleck's interaction, and the results of his calculations are shown together with his experimental results in Table 9.3; remarkably good agreement is achieved.

The examples discussed in this section and in §9.1 show that it is possible to obtain reasonable agreement between theory and experiment. It is, however, important that the experimental crystal should approximate closely to the theoretical model if such agreement is to be expected, and, for the reasons we have given, this state of affairs is usually more readily obtained with rare-earth ions than with transition metal ions.

References

1. Makov, G., Kikuchi, C., Lambe, J. O., and Terhune, R. W., *Phys. Rev.*, 1958, **109**, 1399.
2. Gill, J. C., *Proc. Phys. Soc.*, 1962, **79**, 58.
3. Pace, J. H., Sampson, D. F., and Thorp, J. S., *Proc. Phys. Soc.*, 1960, **76**, 697.
4. Nisida, Y., *J. Phys. Soc., Japan*, 1962, **17**, 1519 and 1965, **20**, 1390.
5. Manenkov, A. A. and Prokhorov, A. M., *Sov. Phys. J.E.T.P.*, 1962, **15**, 54.
6. Donoho, P. L., *Phys. Rev.*, 1964, **133**, A1080.
7. Hemphill, R. B., and Donoho, P. L., *Low Temperature Physics: Proc. 8th Int. Conf. on Low Temp. Phys.* (Butterworths, 1962).
8. Standley, K. J., and Vaughan, R. A., *Phys. Rev.*, 1965, **139**, A1275.
9. Lees, R. A., Moore, W. S., and Standley, K. J., *Proc. Phys. Soc.*, 1967, **91**, 105.
10. Van Vleck, J. A., *Quantum Electronics* (Columbia Univ. Press, 1960).
11. Saunders, D. J., Standley, K. J., and Wilson, P. E., *Brit. J. Appl. Phys.*, 1967, **18**, 1723.
12. Mason, D. R., and Thorp, J. S., *Phys. Rev.*, 1967, **157**, 191.
13. Stevens, K. W. H., and Walsh, D., *J. Phys. C.*, 1968, **1**, 1554.
14. Hoskins, R. H., and Soffer, B. H., *Phys. Rev.*, 1964, **133**, A490.
15. Standley, K. J., and Vaughan, R. A., *Proc. Phys. Soc.*, 1965, **86**, 861.
16. Manenkov, A. A., and Prokhorov, A. M., *Sov. Phys. J.E.T.P.*, 1960, **11**, 527.
17. Davis, C. F., Strandberg, M. W. P., and Kyhl, R. L., *Phys. Rev.*, 1958, **111**, 1268.
18. Armstrong, R. A., and Szabo, A., *Can. J. Phys.*, 1960, **38**, 1304.
19. Mims, W. B., and McGee, J. D., *Phys. Rev.*, 1960, **119**, 1233.
20. Feng, S., and Bloembergen, N., *Phys. Rev.*, 1963, **130**, 531.
21. Orbach, R., *Proc. Roy. Soc.*, 1961, **264**A, 458.
22. Faughnan, B. W., and Strandberg, M. W. P., *J. Phys. Chem. Solids*, 1961, **19**, 155.
23. Scott, P. L., and Jeffries, C. D., *Phys. Rev.*, 1962, **127**, 32.
24. Ruby, R. H., Benoit, H., and Jeffries, C. D., *Phys. Rev.*, 1962, **127**, 51.
25. Cowen, J. A., Kaplan, D. E., and Browne, M. E., *Proceedings of the International Conference on Magnetism and Crystallography, Kyoto, Japan*, 1961.
26. Finn, C. B. P., Orbach, R., and Wolf, W. P., *Proc. Phys. Soc. (London)*, 1961, **77**, 261.
27. Leifson, O. S., and Jeffries, C. D., *Phys. Rev.*, 1961, **122**, 1781.
28. Cowen, J. A., and Kaplan, D. E., *Phys. Rev.*, 1961, **124**, 1098.
29. Huang, Chao-Yuan, *Phys. Rev.*, 1965, **139**, A241.

Appendix 1

COMPARISON OF EXPERIMENTAL METHODS

COMPARISON OF EXPERIMENTAL METHODS

METHOD	ADVANTAGES	DISADVANTAGES
A. Non-resonant (Chapter 5)	Simple circuitry. Independent variation of frequency and magnetic field.	Indirect; interpretation relies on assumption of a physical model. Requires concentrated salts, hence complications due to interionic interactions. Average over all populated levels measured.
B. Continuous saturation (Chapter 6)	Resonant method. Simple circuitry—little modification of conventional spectrometer. No complications due to diffusion within the line, the whole of which is saturated. Cross-relaxation measured by double frequency experiments.	Indirect; interpretation relies on assumption of physical model, and on the values chosen for ancillary parameters (H_1, T_2). Line-shape and sources of broadening require evaluation.
C. Pulse saturation E.S.R. as monitor (§ 7.1)	Resonant method. Direct; measures relaxation parameter of monitored transition. Complicating factors such as cross relaxation, bottlenecks, etc., resolvable by intensive investigation varying experimental parameters. Double-frequency experiments possible.	Unambiguous interpretation of relaxation parameter may require a complete series of measurements. Frequency and field variation not easy.
D. Pulse saturation D.C. magnetization as monitor (§ 7.2)	Spin-flips directly detected. Rudimentary microwave circuit required.	Low sensitivity. Average over all populated levels measured.

Method	Features	Limitations
E. Pulse saturation Faraday rotation as monitor (§ 7.3)	Rudimentary microwave circuit required. Measurement of relaxation possible at fields other than resonant ones. Spatial inhomogeneities in spin temperature detected.	Limited in application since rotation often too small to be measured accurately. Good optical quality of specimens required. Angular variation of relaxation parameter often difficult to interpret. Average over all populated levels measured.
F. Spin-echo (§ 8.1)	Resonant method. T_2 and spin diffusion times often measurable. Sensitivity not limited by monitor signal level. Unambiguous base-line. Useful for broad lines.	Phase-memory time must be greater than pulse duration. Pulse sequence requirements present additional, but not difficult, electronic problem.
G. Inversion recovery (§ 8.2)	Resonant method. Phonon bottleneck detected directly. Hole burning complications obviated by supplying energy across whole line.	Field pulsing required.
H. Optical methods (§ 8.3)	Resonant methods. Phonon bottleneck in Orbach process detected; concentration dependence of this easily investigated. Direct investigation of phonon system possible using Brillouin scattering techniques.	Limited application; requires specific type of energy-level scheme.
I. Linewidths	The only method available at higher temperatures and when $T_1 < 10^{-8}$ sec.	Applicability must be checked by comprehensive line-width studies.

Appendix 2

SUMMARY OF EXPERIMENTAL RESULTS

In the following pages, we have summarized most of the experimental T_1 results published before October 1968. We were greatly helped by the paper *Spin-Lattice Relaxation Time Measurements* by Bates and Malone, which surveyed the literature abstracted before September 1964. We are indebted to those authors and to the Ministry of Defence (Navy Department), under contract from whom that work was carried out, for permission to use information contained in the paper.

We have omitted all results obtained using non-resonant methods. Some early papers giving very sparse experimental details have also been excluded. Other omissions are inadvertent and for these we apologize, as well as for any errors (we hope few) which have arisen in transcribing published data into an abbreviated standard form for presentation.

The Appendix is divided into three sections, transition metal ions, rare-earth metal ions, other ions. In each section, the ions and their host lattices are listed in alphabetical order of their symbols.

The data for each item are given in columnar form in the following order:

A Concentration; any other sample details.

B Experimental method (for identification of letters A to I see column 1 of Appendix 1); frequency in Gc/s (unspecified frequencies in the region of 9–10 Gc/s are referred to as X-band); temperature range in °K.

C Experimental data and theoretical predictions when given.

D Comments.

E Reference (listed numerically at the end of this appendix).

o

TRANSITION METAL IONS

A	B	C	D	E
Co^{2+} in Al_2O_3 Cobalt ions form two non-equivalent magnetic systems (I and II)	B; X-band; 2–60	$H \parallel$ trigonal axis of crystal For I, 9–$30°$K, $T_1 = 1 \cdot 6 \times 10^{-11} \exp(\Delta_I/kT)$ sec with $\Delta_I = 110 \pm 15$ cm^{-1}. Below $4 \cdot 2°$K $T_1 \propto T^{-1}$; value about 1 sec at $4 \cdot 2°$K For II, 14–$26°$K, $T_1 = 10^{-12} \exp(\Delta_{II}/kT)$, $\Delta_{II} = 185 \pm 20$ cm^{-1}. Below $4 \cdot 2°$K, $T_1 \propto T^{-1}$; value about $0 \cdot 1$ sec at $4 \cdot 2°$K		1
Co^{2+} in $La_2Zn_3(NO_3)_{12} \cdot 24H_2O$ $< 0 \cdot 01 \%$	C; $13 \cdot 28$; $1 \cdot 25$	Trigonal site $T_1 \sim 100$–700 msec depending on orientation	Overhauser effect also investigated	2
Co^{2+} in MgO $0 \cdot 003$–$0 \cdot 02\frac{1}{2}\% \ Co^{2+}$, up to $0 \cdot 004\% \ Fe^{2+}$	C and I; 10; $1 \cdot 5$–20 and 50–70	Experimental: $T_1 \sim 1$–5 msec at $4 \cdot 2°$K for different samples Theoretical: $T_1^{-1} = 0 \cdot 1T + 2 \cdot 16 \times 10^{-8}T^9$ $+ 4 \cdot 3 \times 10^{13} \exp(-438/T)$ sec^{-1}	Direct process much faster than theoretically predicted. Raman— very good agreement. Orbach— 100 times slower than theory. Evidence for relaxation via Fe^{2+} was obtained by using specimens with artificially increased amounts of Fe^{2+}/Co^{2+}	3
Co^{2+} in $MgWO_4$ $0 \cdot 01\%$; grown by flux-melt technique	C and I; X-band; $1 \cdot 7$–16 and 30–50	$T_1^{-1} = 220T + 2 \times 10^{-6}T^9$ sec^{-1} from 1–$16°$K $T_1 = 3 \cdot 9 \times 10^{-12} \exp\left(\dfrac{175 \pm 20}{kT}\right)$ sec from 30–$50°$K		4

200

Substance	Concentration		T_1 data	Notes	Ref.
Co^{2+} in TiO$_2$	Concentration not stated	C and I; 9·4; 4·2–23	$H \parallel x$-axis T_1 at 4·2°K = 2 sec 7–23°K, $T_1 = 5\cdot9 \times 10^{-12} \exp(\Delta/kT)$ sec, where $\Delta = 102 \pm 5$ cm^{-1}. Below 7°K this law transforms to $T_1 \propto T^{-1}$		5
Co^{2+} in ZnSiF$_6$. 6H$_2$O	0·01%	C; 13·28; 1·25	Trigonal site $T_1 \sim$ 100–700 msec depending on orientation	Overhauser effect also investigated	2
Cr^{3+} in AlK(SO$_4$)$_2$. 12H$_2$O	0·1–6·0 at. %	C; 9·27; 1·6–4·2	T_1 of order 3 msec at 4·2°K $T_1 \propto T^{-1}$ in this temperature range T_1 concentration independent when $H \parallel$ [110] and [111] but concentration dependence found when $H \parallel$ [100]	Effect of pulse length studied in detail. Van Vleck predicted $T_1 = 1$ msec at 4·2°K and 3000 G. Iron impurity present up to 0·02%	6
	0·36, 1·74, 4·05, 11·7 at. %	C; 9·05; 1·6–4·2	$T_1 = 1\cdot3 \times 10^{-2}\,T^{-1}$ sec for 0·36% sample $T_1 = 1\cdot0 \times 10^{-2}\,T^{-1}$ sec for 1·74% sample $T_1 \propto T^{-1\cdot18}$ for 4·05% sample and $T_1 \propto T^{-1\cdot60}$ for 11·7% sample	Exponential recovery, independent of pulse length, for 11·7% sample. For others, 2 exponentials. Slow decay attributed to T_1	7
Cr^{3+} in Al$_2$O$_3$	10%	B; 9·3; 2–4	$T_1 = 3 \times 10^{-3}\,T^{-1}$ sec	See §9.1 for detailed results	8
Cr^{3+} in Be$_3$Al$_2$Si$_6$O$_{18}$ (emerald)	0·49 × 10^{20} to 1·75 × 10^{20} ions cc	C; 9·3; 1·5–20	$T_1 \sim$ 5 msec at 4·2°K Detailed T_1/orientation curves given	Cross-relaxation processes investigated. Concentration dependence observed for Cr^{3+} > 10^{20} ions/cc	9
Cr^{3+} in CdWO$_4$	$\sim$ 0·03%; grown by Czochralski process	C; 9·4; 1·6–4·2	$H \parallel z$-axis, $\pm\frac{1}{2}$ transition $T_1 = 0\cdot167 \times 10^{-3}\,[\exp(4\cdot85/T) - 1]$ sec $T_1 = 0\cdot36$ msec at 4·2°K $T_1 = 3\cdot0$ msec at 1·6°K	Relaxation via 3/2 level	10

201

A	B	C	D	E
Cr^{3+} in $CsAl(SO_4)_2 \cdot 12H_2O$ 3–100%	B; 9·6; 1·6–4·2	$T_1 \sim 10\ \mu sec$ at 1·6°K No simple temperature dependence	Variation of T_1 with temperature, concentration and transition investigated	11
1–100%	B; 24; 1·6–4·2	$T_1 \propto 1/T$ $T_1 \sim 9$–160 μsec at 4·2°K depending on concentration and transition		12
Cr^{3+} in $K_3Co(CN)_6$ 0·03–1%	G; 9·0; 1·3–4·8	Concentration <0·5%, $T_1 \propto T^{-1\cdot2}$ $T_1 \sim 5$ msec at 4·2°K	Recovery described by single exponential at low concentrations even though a 4-level system. Concentration dependence noted above 0·5%. Hole inverted in inhomogeneous lines	13
0·1%	F; X-band; 4·2	$T_1 = 6$ msec	$T_2 = 10^{-6}$ sec	14
1%	B and modified C; 9·0; 1–4	At 1·3°K, $\pm\frac{1}{2}$ transition T_1 (pulse) $= 3 \times 10^{-2}$ sec T_1 (saturation) $= 4 \times 10^{-2}$ sec	Evidence of phonon bottleneck—probably lattice-bath relaxation observed	104
0·061–0·49% (determined calorimetrically); two non-equivalent complexes	B and C; 9·4; 4·2	$H \parallel c$-axis $T_1 \sim 15$ msec (pulse), independent of concentration $T_1 \sim 15c^{-0\cdot28}$ (saturation) depends on concentration c	Saturation measurements normalized to D.P.P.H. (see Chapter 6). T_2 taken from measured linewidths. Conclude pulse measurements more reliable than saturation	15

Cr³⁺ in K₃Co(CN)₆ 1%, 0·5% and 0·1%	B; 9; 4·2	0·1%; $T_1 = 5 \times 10^{-3}$ sec 0·5%; $T_1 = 2\cdot2 \times 10^{-3}$ sec 1%; $T_1 = 3\cdot5 \times 10^{-5}$ sec	Power at cavity 1×10^{-6} to 5×10^{-2} W. Absolute values probably only correct to order of magnitude. Derivatives normalized using D.P.P.H.	16
Slow evaporation growth method	C; 34·6; 1·4–20	$T_1 \sim 1\cdot2$ msec at 1·4°K, 0·05 msec at 20°K T_1 independent of concentration up to 0·62% Cr		17
0·24%	B and C; 9·4; up to 40	$T_1 \propto T^{-1}$ between 2·1 and 4·2°K $T_1 = 1\cdot9$ msec at 4·2°K	Cross relaxation present. 25 μsec pulses used	18
0·013–0·99%	C; 0·45; 1·5–4·2	$H = 90$ gauss at 18° to c-axis, lower doublet $T_1^{-1} = 320\,c^2$, where c is % concentration, for samples containing up to 0·5% Cr³⁺ Above 0·5%, $T_1^{-1} \sim c^{5\cdot2}$	Possibly relaxation via clusters	19
Cr³⁺ in KAl(SO₄)₂ . 12H₂O 10%	Modified B; X-band; 2–4	$H \parallel (111)$, $\pm\frac{1}{2}$ transition $TT_1 = 3\cdot2 \times 10^{-3}$ sec°K	TT_1 independent of temperature suggests 'direct' process	8
Cr³⁺ in MgO 0·0015%	G; 9·0; 1·3–50	$H \approx 3$ kG $T_1 = 740$ msec at 1·4°K, 300 msec at 4·2°K and 3·8 msec at 50°K	The 'direct' process dominant below 4·2°K. Other E.S.R. lines seen—Fe³⁺, Mn²⁺, V²⁺ and Fe²⁺. Analysis showed impurities up to 10 p.p.m.	20
Cr³⁺ in MgWO₄ Up to 0·23%; grown from melt	C; 9·3; 1·65–4·2	T_1 independent of concentration up to 0·04%; concentration dependence in 0·12 and 0·23% Cr³⁺ samples $H \parallel$ symmetry axis $(-\frac{1}{2}$ to $+\frac{1}{2})$ transition $T_\perp = A[\exp(\Delta/kT) - 1]$, $A = 1\cdot8$ msec, $\Delta = 60$ Gc/s	Saturating pulses 10 μsec to 10 msec long	21

A	B	C	D	E
Cr^{3+} in MgWO$_4$				
0·1 at %; fluxed melt grown	B; X-band; 4·2	$H \parallel z$-axis For $(-\frac{1}{2}$ to $+\frac{1}{2})$ transition, $T_1 = 2\cdot4$ msec For $(-\frac{3}{2}$ to $+\frac{3}{2})$ transition, $T_1 = 0\cdot8$ msec		22
Cr^{3+} in NH$_4$Al(SO$_4$)$_2$. 12H$_2$O				
2 at. %	C; 8·75; 4·2	$T_1 = 7\cdot5 - 9$ msec		23
Cr^{3+} in TiO$_2$				
0·07%	C; 34·6; up to 20	$H \perp c$-axis $T_1 = 3-6$ msec at $1\cdot4°$K, $2-4$ msec at $4\cdot2°$K, $1\cdot4-2$ msec at $10\cdot1°$K, $0\cdot6$ msec at $20°$K; varies between transitions		24
4–300 p.p.m.; Fe 40–90 p.p.m.	G; 1·7 and 3·5; 1·2–4·2	1–2 and 3–4 transitions measured At low temperatures, T_1 behaviour similar to Orbach type $T_1 \sim 150$ msec at $1\cdot2°$K, 20 msec at $4\cdot2°$K	Monitor power level kept to $< 10^{-8}$ W to avoid saturation effects. Detailed theoretical discussion, agreement with experiment not good. Three decay times (four-level system) expressed in terms of two temperature independent parameters	25
0·1%	C; 9·4; 1·7–4·2	$H \parallel a$-axis T_1 depends on transition; $2-4$ msec at $4\cdot2°$K and $3\cdot3 - 9$ msec at $1\cdot7°$K	Cross-relaxation effects observed	26
Cr^{3+} in ZnWO$_4$				
~ 0·1 to 0·5% in melt; grown by Czochralski technique	C; X-band; 1·8–4·2	$T_1 = A \exp(\Delta/kT)$ $A = 0\cdot15$ msec, $\Delta = 100$ Gc/s	Separation of $+\frac{3}{2}$ and $-\frac{1}{2}$ levels is calculated to be 95 ± 5 Gc/s	27
Up to 0·27%; grown by Czochralski technique	C; 9·3; 1·65–4·2	T_1 independent of concentration below 0·1% $T_1 = A[\exp (\Delta/kT) - 1]$, $A = 1\cdot0$ msec, $\Delta = 70$ Gc/s		21

Cr³⁺ in ZnWO₄

Sample	Conditions	Results	Notes	Ref.
Up to 0·8%; Czochralski technique used. Care taken to ensure proper charge compensation, using Li^+ in the melt	C; 9·2 and 33·8; 1·6–20	9·2 Gc/s: For lower doublet T_1 independent of Cr^{3+} concentration up to 0·3 at.% and $\sim$ 10 msec at 1·5°K. For higher concentrations T_1 roughly $\propto$ (conc.)$^{-1}$. Orbach temperature dependence, $\Delta \sim$ 50 Gc/s 33·8 Gc/s: $T_1 \sim$ 2·5 msec at 1·5°K, Orbach temperature dependence with $\Delta \sim$ 43 Gc/s, plus significant 'direct' term at $T =$ 1·5°K	Pulse length (9·2 Gc/s) 5 μsec, and 10–100 μsec (33·8 GHz). Results fit theory well except for frequency dependence	28
0·1%; fluxgrown	C; 9·4; 1·6–4·2	$H \parallel z$-axis $T_1 =$ 1·1 msec at 4·2 and 5·3 msec at 1·6°K $T_1 =$ 1·15 [exp(Δ/kT) $-$ 1] 10^{-3} sec $\Delta/k =$ 2·8°K		29

Cu²⁺ in copper (α α' Br) dipyrromethene

Sample	Conditions	Results	Notes	Ref.
Grown from manufacturers 'pure' materials. 100%	C; 9·4; 1·6–4·2	$T_1^{-1} =$ 82T^2 + 3·3 × 10^{-2} T^9 sec^{-1} $T_1^{-1} =$ 36T^2 + 1·65 × 10^{-3} T^9 sec^{-1}, for two samples	Phonon bottleneck investigated. Nearest Cu–Cu distance 10 Å in stoichiometric crystal. Very small (undetected) h.f.s. No detectable impurity resonances	30

Cu²⁺ in copper phthalocyanine

Sample	Conditions	Results	Notes	Ref.
Grown from acid purified powdered material, impurities unknown. 100%	C; 9·4; 1·6–4·2	Three samples, $1/T_1$ represented by (a) 68T + 11·8 × 10^{-4} T^9 sec^{-1} (b) 63$T^{1·38}$ + 8·9 × 10^{-4} T^9 sec^{-1} (c) 102·5$T^{1·18}$ + 7·3 × 10^{-4} T^9 sec^{-1}	No detectable impurity resonance. Behaviour of 'direct' term not explained	30

Cu²⁺ in [(C₄H₉)₄N]₂[S₂C₂(CN)₂]₂Cu

Sample	Conditions	Results	Notes	Ref.
100%	C; 9·27; 1·7–4·2	In symmetry plane $T_1^{-1} \propto T^2$ $T_1 \sim$ 200–600 msec at 4·2°K $T_1 \sim$ 2–3·5 sec at 1·7°K for different samples	Effects of variation of monitor power level investigated. Quoted results required $<$ 10^{-8} W at cavity for consistency. Results fitted to bottleneck theory (see p. 70)	31

A	B	C	D	E
Cu^{2+} in $La_2Mg_3(NO_3)_{12} \cdot 24H_2O$ and its deuterated counterpart Very pure starting materials (99·997%)	C; 9·4; 1·3–20	For Cu concentrations about 5×10^{17} ions/cc and deuterated samples, $T_1^{-1} \approx 100T + 3·3 \times 10^{-2} T^5$ For protonated sample, Cu concentration 10^{18} ions/cc, $T_1^{-1} \approx 13 \times 100 T$ in range 1·3–10°K Some dependence of T_1 on Cu concentration and on orientation	Dynamic Jahn–Teller system. Only detected paramagnetic impurity, 1·5 p.p.m. Fe. Signal recovery independent of pulse length, and monitor power. The low temperature rate is four orders of magnitude faster than that observed for Cu^{2+} in the static octahedral water co-ordination in the KZn Tutton salt	32
Cu^{2+} in $La_2Mg_3(NO_3)_{12} \cdot 24H_2O$ 0·1% (nominal)	C; X-band; 1·6–4·2	Two time constants fitted to recovery $T_1^{-1} = 2300T$ sec^{-1} (initial recovery) $T_1^{-1} = 250T^2$ sec^{-1} (final part of recovery curve)	After 10 msec wide pulse, recovery non-exponential. The very short relaxation times are explained in terms of strong coupling to non-conjugate states	33
Cu^{2+} in $Cu(NH_4)_2(SO_4)_2 \cdot 6H_2O$	C; X-band; 1·34–4·2	Size dependence of T_1 found. For crystal of length L, at 2·09°K $T_1 = 100$ msec, $L = 150\mu$ $T_1 = 350$ msec, $L = 600\mu$		34
Cu^{2+} in $K_2Zn(SO_4)_2 \cdot 6H_2O$ 0·1 – 5·0 at. %	C; 9·5; 1·2–14	$H \parallel (1\bar{1}0)$ plane For the most dilute samples at temperatures up to 4·2°K, $T_1^{-1} = \alpha T + \beta T^9$, with $\alpha = (2·1 \pm 0·1) \times 10^{-2}$ sec^{-1}°K^{-1} and $\beta = (1·9 \pm 0·2) \times 10^{-6}$ sec^{-1}°K^{-9} Relaxation rate less rapid than this equation for $T > 10$°K Concentration dependence of T_1 observed	Iron impurity present, not more than 0·01 at. %, but no ferric iron resonance detected. Theory given in reference 36	35

206

Cu²⁺ in Zn(NH₄)₂(SO₄)₂ . 6H₂O Cu^{2+} in $Zn(NH_4)_2(SO_4)_2 . 6H_2O$ $0\cdot02$–100%	C; $9\cdot0$; $1\cdot34$–$4\cdot2$	$T_1 \sim 20$ msec at $4\cdot2°K$ (12% and 25% specimens)	Measurements performed as a function of concentration and crystal size. Linear dependence on crystal size noted at higher concentrations. Deuterated salt gave order of magnitude different results at 1%. Results depended on speed of growth of specimens	37
1%	B and modified C; $9\cdot0$; $1\cdot4$	At $1\cdot3°K$ on the $\pm\frac{1}{2}$ and $\pm\frac{3}{2}$ transitions T_1 (pulse) $= 20$ sec T_1 (saturation) $= 2$ sec	Method B unreliable because of low saturating power, and T_2 values from linewidth measurements are suspect. Cross saturation between the eight hyperfine lines investigated.	104
Cu²⁺ in Zn(BrO₃)₂ . 6D₂O Cu^{2+} in $Zn(BrO_3)_2 . 6D_2O$ (i) 7×10^{17} ions/cc (ii) 30×10^{17} ions/cc	C; $9\cdot4$; up to 20	$T_1^{-1} = 7\times10^3 T$ sec^{-1} from $2\cdot2$ to $4\cdot2°K$ for (i) $T_1^{-1} = 9\cdot5\times10^3 T$ sec^{-1} from $2\cdot2$ to $9°K$ for (ii)	No E.S.R. detected in undoped bromate samples. Some Fe detected spectrographically. Jahn–Teller system	32
Cu²⁺ in ZnWO₄ Cu^{2+} in $ZnWO_4$ up to $0\cdot5\%$ Cu^{2+} $< 0\cdot01\%$ Fe	C; $9\cdot27$; $1\cdot6$–$4\cdot2$	For samples $< 0\cdot05\%$ T_1^{-1} (x-axis) $= 2T + (0\cdot37T)^9$ sec^{-1} T_1^{-1} (y-axis) $= 2\cdot7T + (0\cdot36T)^9$ sec^{-1} T_1^{-1} (z-axis) $= 0\cdot35T + (0\cdot33T)^9$ sec^{-1}	Angular variation investigated. Concentration dependence observed at the higher levels of doping, approaching a T^2 temperature dependence attributed to relaxation via pairs	38
Fe²⁺ in MgO Fe^{2+} in MgO $0\cdot02\%$ Fe^{2+} $0\cdot05\%$ Fe^{3+} $0\cdot002\%$ Cr^{3+} $0\cdot002\%$ Mn^{2+}	C; $9\cdot15$; $1\cdot8$–7	At $4\cdot2°K$ $H \parallel (001)$ Experimental Theoretical $\tau_a T$(sec °K) $1\cdot0\times10^{-4}$ $0\cdot72\times10^{-4}$ $\tau_b T$(sec °K) $2\cdot6\times10^{-4}$ $4\cdot5\times10^{-4}$	Three-level system, hence two relaxation times τ_a and τ_b. Theoretical values calculated from parameters obtained from acoustic experiments. Measured angular dependence satisfactorily predicted	39

A	B	C	D	E
Fe²⁺ in MgO 0·06% Fe²⁺ 0·004% Mn²⁺ 0·0004% Cr³⁺	C; 9·2; 3–50	$T_1^{-1} = 4 \times 10^3 T + 7·9 \times 10^{-3} T^7 + 2·8 \times 10^{13} \exp(-200/T) \sec^{-1}$		40
		$T_1^{-1} = 4 \times 10^3 T + 5 \times 10^{-3} T^7 + 2·9 \times 10^{11} \exp(-150/T) \sec^{-1}$	Correction to earlier publication	41
Fe³⁺ in Al₂O₃ 0·03%	C; 34·6; up to 20	$T_1 \propto T^{-1}$ below 10°K. $T_1 \approx 4–13$ msec at 1·4°K, 2·0 msec at 4·2°K, 0·8 msec at 10·1°K and 0·5 msec at 20°K. Varies between transitions		24
0·02%	C, B, and I; X-band; 2–80°K	2–5°K, $T_1 \propto T^{-1}$; 5–15°K, small change only in T_1; 20–80°K, $T_1 \propto T^{-6}$ approx.	Concentration dependence and cross relaxation observed	42
Fe³⁺ in Al(NH₄)(SO₄)₂ . 12H₂O 0·055–0·012%	Modified B; X-band; 2–4	$H \parallel (100)$ $TT_1 \sim 5 \times 10^{-4} \sec °K$ for $\pm\frac{1}{2}$ transition $TT_1 \sim 14 \times 10^{-3} \sec °K$ for $(\frac{3}{2}$ to $\frac{1}{2})$, $(-\frac{3}{2}$ to $-\frac{1}{2})$ transitions	Temperature independence of TT_1 suggests direct process. No concentration dependence observed	8
Fe³⁺ in RbAl(SO₄)₂ . 12H₂O 0·3–1·1%; ground state $S = \frac{5}{2}$	C; 9·4; 1·6–4·2	$T_1 \propto T^{-1·5}$ below 1%, $T_1 \sim 1$ msec at 4·2°K	Measurements made on well resolved central line of complex spectrum. Traces non-exponential, and marked angular dependence observed due to cross relaxation. T_1 concentration dependent above 1%	43

Fe^{3+} in K$_3$Co(CN)$_6$

Concentration	Conditions	T_1 data	Comments	Ref.
0·1–3·0 at. %	C; 8·75; 1·6–4·2	$T_1^{-1} = 5\cdot4T + 5\cdot4\times10^{-3}\,T^9$ sec^{-1} $T_1 \propto 1/T$ below 1·8°K Theory shown to yield $T_1^{-1} \approx 15T + 0\cdot1T^9$ sec^{-1}	500 mW pulse power. Little concentration dependence	44
0·1 and 0·2 at. %	G; 2–4 and 8–12; 1·2–4·2	$H \parallel c$-axis $T_1^{-1} = 46\cdot5T + 4\cdot95\times10^{-3}\,T^9$ sec^{-1} $T_1 \sim$ 10 msec at 2·15°K and 9·24 Gc/s	Recovery curves non-exponential at high frequencies or for large samples, suggestive of bottleneck. Direct term frequency dependence checked ($\nu^2 H^2$). Theory given in reference 46	45
0·2 and 0·44%	B and C; 9·4; 4·2	$H \parallel b$-axis $T_1 \sim$ 1 msec, concentration dependent	Saturation measurements normalized to D.P.P.H. (see Chapter 6). T_2 taken from linewidths. Accuracy of pulse data doubted by authors	15
Nominal 0·1–5%; ground state has $S = \frac{1}{2}$	C; 9·4; 1·6–4·2	$H \parallel b$-axis Experimental: $T_1^{-1} = 15T + 0\cdot02T^8$ sec^{-1}	No concentration dependence observed	43
1·7, 3·2 and 6·6% Grown by slow evaporation	C; 9·375; 1·6–4·2	Below 2·8°K $\quad T_1 \propto 1/T$ Above 2·8°K $\quad T_1 \propto T^{-7}$	Direct term slightly concentration dependent	47
0·1 and 0·21%	B and C; 9·4; up to 40	T_1 independent of concentration from 2·1–40°K and $T_1 \approx 64T^{-5}$ sec $T_1 = (1\cdot6\pm0\cdot2)\times10^{-2}$ sec at 2·1°K $T_1 = (4\cdot9\pm0\cdot4)\times10^{-4}$ sec at 4·2°K	25 μsec pulses, 1·5 W peak power	18
0·21%	C; 3·17; 1·77 and 4·2	$T_1^{-1} = 0\cdot093T + 0\cdot0054T^9$ sec^{-1} at this frequency agreeing with $T_1^{-1} = 5\cdot4T + 0\cdot0054T^9$ sec^{-1} at 8·75 Gc/s due to previous measurements (reference 44)	Van Vleck theory used to show agreement	48

A	B	C	D	E
Fe^{3+} in $K_3Co(CN)_6$ 0·1, 0·21, 0·46%	B; 42 Mc/s; 0·08–4·2	Most measurements made with $H \parallel c$-axis. 2–4°K, $T_1 \propto T^{-9}$ for 0·1% sample. At 4·2°K, $T_1 \sim 10^{-3}$ sec. Below 2°K, Only at lowest temperatures $T_1 \propto T^{-1}$; For 0·21% sample region of constant T_1 from 0·5–2°K for $H \parallel c$-axis, not found along a- or b-axes	Theoretical discussion of results in comparison with 10 Gc/s data	49
0·24–3·5 at. %	G; 1·8 and 8·5; 1·25–4·5	For 0·24% sample $T_1^{-1} = 4·2 \times 10^{-3} T^9 \text{ sec}^{-1}$ at 1·8 Gc/s $T_1^{-1} = 3·1T + 4·3 \times 10^{-3} T^9 \text{ sec}^{-1}$ at 8·5 Gc/s $T_1 \sim 460$–600 μsec at 4·2°K for low concentrations	Frequency dependence of direct term, and concentration dependence of T_1 observed. Monitor power < 1 μW	50
Fe^{3+} in TiO_2 Up to 2×10^{19} ions/cc	C; 8·55; 1·4	Four transitions all give $T_1 \sim 4$ msec at 1·4°K One measurement indicated 0·1 msec at 78°K		51
Concentration 'sufficiently dilute for there to be negligible cross relaxation'. Ground state $S = \frac{5}{2}$	C; 9·4; 1·6–4·2	$T_1 \sim 1·6$–$2·2$ msec at 4·2°K $T_1 \sim 4·2$–5 msec at 1·7°K $T_1 \propto T^{-1}$ along axes	Two equivalent magnetic complexes per unit cell with their local symmetry axes at right angles to one another	43
Concentration not given	C; 9·4; 1·7–4·2	$H \parallel [110]$ $T_1 \sim 2$ msec at 4·2°K and 6 msec at 1·7°K	Cross-relaxation effects observed	26
Fe^{3+} in $ZnWO_4$ 0·1 and 0·3%; fluxgrown	C; 9·4; 1·6–4·2	0·1%, $T_1 \propto T^{-1}$ and $T_1 \sim 75$ μsec at 1·6°K 0·3%, $T_1 \sim 85$ μsec at 1·6°K		29

Fe^{3+} ZnWO$_4$ 0·08%	C; X-band; 1·6–4·2	$T_1 \sim 0.3$ and 1 msec for upper and lower doublet respectively at 4·2°K. For lower doublet, $T_1 \propto [\exp(\Delta/RT) - 1]$ with $\Delta \sim 90$ Gc/s	Exploratory measurement only	52
Mn^{2+} in BaF$_2$ 0·004–0·03%; grown by Stockbarger technique	B and G; 9; 1·2–11	$T_1^{-1} = 118T + 8 \times 10^{-1}\,T^5\ \text{sec}^{-1}$	No angular dependence. Iron present in similar concentrations	53
Mn^{2+} in CaCO$_3$ 2 × 10^{18} ions/cc	F; X-band; 4·2 and 77	$T_1 = 3$ msec at 4·2°K	$T_2 = 3$ μsec at 4·2°K, temperature dependent	14
Mn^{2+} in CaF$_2$	B and G; 9·47; 2·5–300	Spectrum I $T_1^{-1} = 25T + 3·8 \times 10^{-5}\,T^5\ \text{sec}^{-1}$ Spectrum II $T_1^{-1} = 19T + 4·5 \times 10^{-4}\,T^{3·6}\ \text{sec}^{-1}$	Saturation measurements of T_1 not absolute, but normalized to those obtained by the rapid passage method where these overlap. Spectrum I is the normal one; Spectrum II is produced by rapid cooling from ∼ 700°C	54
Mn^{2+} in CaWO$_4$ 3 × 10^{16} ions/cc Paramagnetic ions substituted on Ca^{2+} sites	B; X-band; 100–300	$T_1 \propto T^{-3/2}$ $T_1 \sim 6·6 \times 10^{-8}$ sec at 130°K	T_2 estimated from linewidth of F-h.f.s. line	55
	G; 9·4; 8–26	$\pm\frac{1}{2}$ transition, $H \parallel c$-axis $T_1^{-1} \propto T^{4·8}$	Thirty allowed E.S.R. transitions. Approximately a T^5 law expected for a multilevel system. Primarily an investigation of phase memory times in spin echoes	56
Mn^{2+} in CdF$_2$ 0·01%	C; 9·3; 1·6–4·2	T_1 varies with transition and orientation. $T_1 \propto T^{-1}$, $T_1 \sim 10$ msec at 4·2°K		57
Mn^{2+} in MgO 0·01%	C; 9·3; 1·3–4·2	$T_1T = 2$ sec °K. For ionic concentration of 0·01%, cross-relaxation time = 100 msec		58

A	B	C	D	E
Mn²⁺ in MgO				
0·01%, powdered sample	C; 9·3; 1·6–4·2	$T_1 = 0\cdot014T^{-1}$ sec		57
Up to 1900 p.p.m. Mn²⁺ 40 p.p.m. Fe³⁺	C; 9·6; 1·35–4·2	$T_1 = 2$ sec for 0·18 p.p.m. Mn²⁺ $T_1 = 4$ msec for 2000 p.p.m. Mn²⁺ at 4·2°K About 50% increase in T_1 in cooling from 4·2–1·5°K	Cross-saturation experiments performed. T_1 sensitive to sample treatment. No obvious systematic variation with concentration	59
Mn²⁺ in MgSO₄ . 4H₂O				
50%	B; 9·3; 1·9–4·2	$T_1 = 25$ μsec at 4°K	T_1 independent of temperature below 2·5°K; attributed to cross relaxation	60
Mn²⁺ in NaF				
~10¹⁸ and 10¹⁹ ions/cc	B; 9; 2–100	$T_1^{-1} = 1\cdot33 \times 10^3 T + 1\cdot2 \times 10^{-12} T^9 \text{ sec}^{-1}$		16
Mn²⁺ in SrF₂				
0·004%; grown by Stockbarger technique	B and G; 9; 1·2–12	$T_1^{-1} = 154T + 5 \times 10^{-2} T^5 \text{ sec}^{-1}$	Unidentified impurity detected; also iron present in similar concentration	53
Mn²⁺ in SrS				
0·05%; polycrystalline specimens	B, C and I; 9·3; 1·6–300	300°K (linewidths) $T_1 \sim 5 \times 10^{-8}$ sec 77°K (saturation) $T_1 \sim 1\cdot5 \times 10^{-6}$ sec 4·2°K (saturation) $T_1 \sim 2\cdot9 \times 10^{-2}$ sec 4·2°K (pulse) $T_1 \sim 9 \times 10^{-2}$ sec	Cross relaxation observed. T_2 taken as 6×10^{-8} sec from observed linewidth	62
0·05%, powdered sample	B, C and I; 9·3; 1·6–4·2, 300	$T_1 = 5 \times 10^{-9}$ sec at 300°K, $1\cdot5 \times 10^{-6}$ sec at 77°K, 200 msec at 4·2°K	T_1 at helium temperatures found to vary with concentration	57
Mn⁴⁺ in Al₂O₃				
	H; 24 and 48; 6–9	$T_1 = C \exp(\Delta/kT)$ sec, where $C = 1\cdot6 \times 10^{-10}$ sec and $\Delta = 72 \pm 20 \text{ cm}^{-1}$	Relaxation in $\bar{E}(^2E)$ state	63

212

Ion and host	Code	Relaxation data	Comments	Ref.
Ni^{2+} in MgO	C; 9·15; 1·8–7	$H \parallel [001]$ at 4·2°K Experimental: $\tau_b T = 2·6 \times 10^{-3}$ sec °K, $\tau_b T = 3·4 \times 10^{-3}$ sec °K Theoretical: $\tau_a T = 8·9 \times 10^{-3}$ sec °K, $\tau_b T = 4·3 \times 10^{-2}$ sec °K	Relaxation of three level system governed by two time constants τ_a and τ_b. Poor agreement between theory and experiment possibly due to impurities	39
0·007% Ni^{2+} 0·03% Fe^{2+} 0·002% Fe^{3+} 0·004% Mn^{2+} 0·001% Cr^{3+}	C; 9·2; 3–50	$T_1^{-1} = 2·45 \times 10^2 T + 8 \times 10^{-7} T^7 + 1·5 \times 10^2$ sec^{-1}	The constant term in T_1^{-1} is attributed to cross relaxation	40
Ni^{2+} in ZnSiF$_6$. 6H$_2$O 0·02–100 at. %	C; 3·8 and 4·4; 1·8–4·2	Temperature dependences range from $T^{1·6}$ to T^3 between 2 and 3°K and from $T^{2·2}$ to T^4 between 3 and 4°K. Concentration effect observed ~ 1 msec at 4·2°K	Pulse power 0·2 W, duration 1 μsec to 1 msec. Monitor power 1 μW	64
0·05–1·0%	C; 9·4; 1·5–4·2	At 1·5°K, $T_1 = 1·9$ msec (0·05% sample) and 7·1 msec (1% sample) $T_1 \propto T^{-1·7}$ for 1% sample	Phonon bottleneck suspected	65
Ni^{3+} in Al$_2$O$_3$	I; 9·3; 77–500	$T_1 = (0·6 \pm 0·3) \times 10^{-11} \times \exp[(1530 \pm 150)/kT]$ sec	Jahn–Teller splitting of ^2E ground state	66
Ni^{3+} in SrTiO$_3$	I; 9·3; 77–500	$T_1 = (20 \pm 5) \times 10^{-11} \exp[665 \pm 50)/kT]$ sec	Jahn–Teller splitting of ^2E ground state	66
Ti^{3+} in Al$_2$O$_3$	C; 9·3; 1·7–3·5	$H \parallel$ threefold crystal axis. $T_1 \propto \exp(\Delta/kT)$, $\Delta = 30 \pm 3$ cm^{-1} and $T_1 \sim 1$ msec at 2°K. Angular dependence measured.		67

A	B	C	D	E
Ti^{3+} in Al_2O_3	B and I; X-band; below $9°K$	$H \parallel c$-axis $T_1 \sim 5 \times 10^{-8}$ sec at $9°K$, 10^{-4} sec at $4\cdot3°K$ and 10^{-1} sec at $1\cdot55°K$		68
V^{2+} in Al_2O_3	H; 24 and 48; $1\cdot4$-$2\cdot15$	$T_1 = C \exp(\Delta/kT)$ sec, where $C = 5\cdot2 \times 10^{-8}$ sec and $\Delta = 12\cdot1 \pm 1\cdot2$ cm^{-1}	Relaxation in $\bar{E}(^2E)$ state. No frequency dependence. Reasonable fit also to $T_1 \propto T^{-9}$, but optically measured value of $\Delta \sim 12\cdot3$ cm^{-1}	63
V^{4+} in TiO_2 $0\cdot05\%$	F; 9; $1\cdot3$-$4\cdot2$	Two recovery parameters of value $0\cdot1$ and $0\cdot67$ sec respectively and *independent* of temperature in this range. Independent of 15% change in sample size and of surface condition. Dependent on annealing history	Pulse power 1 kW, pulse width ~ 30 nsec. Time between pulses 3 μsec.	69
$0\cdot01\%$	C and I; $9\cdot4$; $4\cdot2$-110	$4\cdot2$-$10°K$, $T_1 \propto T^{-1}$ 10-$50°K$, $T_1 \propto T^{-5\cdot5}$ $70°K$, $T_1 = 6 \times 10^{-12} \exp(\Delta/kT)$ sec, where $\Delta = 650$ cm^{-1} $4\cdot2°K$, $T_1 \sim 20$ msec	Saturating pulse widths 10 μsec- 1 msec	70

A	B	C	D	E
Ce^{3+} in CaF$_2$ 0.08 to 1.6 % by wt.; cerium in tetragonal sites	C; 9.6; 2–15	H applied in (100) plane Below 10°K, T_1 concentration dependent $T_1 = 3.2 \times 10^{-3} T^{-1}$ sec for 0.08% sample $T_1 = 3.1 \times 10^{-4} T^{-1}$ sec for 0.34% sample Above 10°K, results for different samples converge to $T_1 = 4 \times 10^5 T^{-9}$ sec	Angular variation of T_1 observed	71
	C; 9.27; 1.6–4.2	$H \parallel z$-axis and $H \perp z$-axis $T_1 \propto T^{-1}$ $T_1 = 170$ μsec at 1.6°K $T_1 \sim 70$ μsec at 4.2°K		72
Ce^{3+} in CaWO$_4$ 0.003%	F; 9.4; 1.4–9	$H \parallel c$-axis $T_1^{-1} = 0.14T + BT^{10.9}$ sec^{-1}		73
	C; 9.27; 1.6–4.2	At 4.2°K, $T_1 \sim 200$ msec At 1.6°K, $T_1 = 2.4 \pm 0.2$ sec Temperature dependence faster than T^{-1}		72
Ce^{3+} in La(C$_2$H$_5$SO$_4$)$_3$. 9H$_2$O 0.2%	C; 9.36; 1.4–1.8	$H \perp z$-axis Experimental: $T_1^{-1} = 2.2 \times 10^6 \exp(-5.67/kT)$ sec^{-1} Theoretical: $T_1^{-1} = 6.9 \times 10^7 \exp(-5.67/kT)$ sec^{-1}	Discrepancy possibly due to phonon bottleneck	74
Ce^{3+} in LaCl$_3$ 0.1–2%	C; 9.5; 1–4.2	$T_1 \sim 100$ μsec at 3.6°K $T_1 \sim 700$ msec at 2°K $T_1 \sim 2$ sec at 1.1°K Orbach process above 2°K, $\Delta/k \sim 46$°K	Exponential time base used to produce linear traces on an oscilloscope. Decays non-exponential. Direct process dominates below 2°K	75

215

A	B	C	D	E
Ce³⁺ in LaCl₃ 1 and 2%	C; 9·15; 1·65–4·2	$H \parallel c$-axis $T_1^{-1} = 0.54T + 2.6 \times 10^9 \exp(-46/T)$ sec⁻¹	Angular variation in 'direct' process fitted to theory	76
Ce³⁺ in LaF₃ 1%	C; 9·3; 1·3–5	$H \parallel c$-axis $T_1^{-1} = 0.41T + 5.8 \times 10^8 \exp(-56/T) + 2.1 \times 10^{-4} T^9$ sec⁻¹ $H \parallel y$-axis $T_1^{-1} = 33 \operatorname{cosech}(3.7/T) + 2.3 \times 10^{-3} T^9$ sec⁻¹	Slight concentration dependence of 'direct' term noticed. Orbach-like term possibly due to cross relaxation to ~ 0·01% Nd³⁺ impurities, since no excited level at 56 cm⁻¹ for Ce³⁺	77
Ce³⁺ in La₂Mg₃(NO₃)₁₂ · 24H₂O				
0·2 and 0·5%	G; 1·7–11·14; 1·25–4·2	$T_1^{-1} = 6.02 \times 10^{-39} v^4 T + 0.09 T^9 + \exp(\log B - \Delta/T)$ sec⁻¹, where v in c/s, T in °K, $\Delta = 36.7$°K	Some variation in parameters with frequency	78
0·2 and 0·5%	G; 11·35 and 11·45; 1·2–4·2	Experimental: $$T_1^{-1} = \left(\frac{1}{6.02 \times 10^{-39} v^4 T + 0.09 T^9 + E \exp(-36.7/T)} + \frac{1}{DT^2} \right)^{-1}$$ where D and E vary from sample to sample over range 1·25–2·5°K	Monitor power $< 10^{-7}$ W. Found T_1 depended on concentration, surface condition and size. Bottleneck observed; $\sigma \sim 5/T$	79
0·2%	F; 9·3; 1·8–4	$T_1 = 6 \times 10^{-10} \exp(34/T)$ sec $T_1 \sim 1$ μsec at 4·2°K	3 pulse sequence. 100 W, 50 nsec pulses. T_2 also measured	80
0·2%	F; 9·3; 1·5–4·2	$T_1 = 7 \times 10^{-10} \exp[(34 \pm 2)/T]$ $T_1 \sim 10^{-4}$ sec at 2·5°K		81
0·2%	F; X-band; 1·5–4·2	$T_1 = 9 \times 10^{-7}$ sec at 4·2°K $T_1 = 40 \times 10^{-3}$ sec at 1·5°K	$T_2 = 2 \times 10^{-7}$ sec at 4·2°K, temperature dependent. Measured T_1 values consistent with $T_1 \propto (34/T)$	82

Ce³⁺ in La₂Mg₃(NO₃)₁₂ · 24H₂O 0·05–100%	C; 9·4; 1·7–2·7	$T_1 \propto T^{-14}$ down to 1·9°K or $T_1 \propto \exp[(32\pm2)/T]$	Monitor power $\sim 10^{-5}$ W	83
0·2 and 2%	C; 9·6; 0·25–1·8	$H \perp z$-axis Experimental: $T_1^{-1} = D \coth^2(h\nu/2kT)$ sec⁻¹, where $D=0·8$ sec⁻¹ for the 2% sample, and 2·4 sec⁻¹ for the 0·2% sample Theoretical: $T_1^{-1} = 1·3 \coth(h\nu/2kT)$ $+3·5 \times 10^9 \exp(-34/T)$ sec⁻¹		84
Ce³⁺ in Y(C₂H₅SO₄)₃ · 9H₂O 2%	C; 9·36; 1·2–2·5	H at 65° to c-axis Experimental: $T_1^{-1} = 5·1 \times 10^8 \exp(-25/T)$ $+1·0 T^9$ sec⁻¹ Theoretical: $T_1^{-1} = 0·048T + 4·1$ $\times 10^8 \exp(-25/T) + 1·8 \times 10^{-4} T^9$ or $3·39T + 3·2 \times 10^{10} \exp(-25/T) + 1·0 T^9$	Experimental value of Δ/k accurate to ±1°K. Predicted value of T_1 depends on approximations in theory	85
Ce³⁺ in YCl₃ · 6H₂O 0·1 and 1%	C; 9·36; 1·3–5	$H \parallel z$-axis, 1% sample $T_1^{-1} = 0·27T + 2·5 \times 10^{-5} T^9$ $+2 \times 10^7 \exp(-60/T)$ sec⁻¹	Angular dependence measured for 0·1% sample at 1·4°K	86
Dy³⁺ in CaF₂ 0·1–0·2%	C; 9·6; 2–4	No measurable recoveries observed Upper limit of T_1 estimated as 3 μsec		71
Dy³⁺ in LaF₃ 0·5–1·5%	C; 9·3; 1·3–5	$H \parallel x$-axis $T_1^{-1} = AT + 6·5T^9$ sec⁻¹ $A \sim 200$ sec⁻¹ deg K⁻¹, depending on concentration	Non-uniform variation of A with concentration suggests dependence owing to cross relaxation to pairs. Probably not the true T_1 which is measured	77
Dy³⁺ in La₂Mg₃(NO₃)₁₂ · 24H₂O $\sim 1\%$	C; 9·35; 1·2–1·8	$H \parallel z$-axis Experimental: $T_1^{-1} = 7·0 \times 10^9$ $\exp(-22·0/T)$ Theoretical: $T_1^{-1} = (<2T) + 3·1$ $\times 10^9 \exp(-22·/T) + (\sim 0·2T^9)$	Crystal grown from 50% solution	85

A	B	C	D	E
Dy³⁺ in Y₃Al₅O₁₂ Nominal 1%	C; 8·7; up to 20	$T_1^{-1} = 10T + 2\cdot4 \times 10^9 \exp(-53 \times 1\cdot44/T)\,\text{sec}^{-1}$	No optical measurements of splitting available	87
Dy³⁺ in YCl₃ . 6H₂O 0·1–5 at. %	C; 9·36; 1·3–5	$H \parallel z\text{-axis}$ Experimental: $T_1^{-1} = 2\cdot25T + 5 \times 10^{-3}T^9$ $+8 \times 10^9 \exp(-49/T)\,\text{sec}^{-1}$ $H \parallel x\text{-axis}$ $T_1^{-1} = \dfrac{225 \times 250T^2}{225T+250} + 5 \times 10^{-3}T^9 + 8$ $\times 10^9 \exp(-49/T)\,\text{sec}^{-1}$	Large anisotropy observed at 1·4°K. Possibly a bottleneck for the higher concentration specimens for $H \parallel x$-axis	86
Dy³⁺ in Y(C₂H₅SO₄)₃ . 9H₂O 0·1%	C and I; 9·5; 1·4–20	$T_1 = 3 \times 10^{-9}[\exp(19\cdot8/T) - 1]\,\text{sec}$	Measurements made also on first excited doublet	88
Dy³⁺ in Y₃Ga₅O₁₂ Nominal 1%	C; 8·7; up to 20	$T_1^{-1} = 440T + 11 \times 10^7$ $\exp(-14 \times 1\cdot44/T)\,\text{sec}^{-1}$	No optical measurements of splitting available. Some indication that impurities affected the results	87
Er³⁺ in BaF₂ 0·01–0·2%	C and I; 9·4 and 36; 1·4–30	$T_1^{-1} = 115 \coth(0\cdot863/T) + 8 \times 10^{-5}T^9$ $+ 1\cdot4 \times 10^{11} \exp(-107/T)\,\text{sec}^{-1}$ at 36 Gc/s	'Direct' term varies between samples of different concentration, and is $2\cdot6T$ at 9·4 Gc/s for the 0·05% sample	89
~ 0·1%	C and I; 9 and 36; 1·6–25	$T_1 = 10^{-11} \exp(70/kT)$ sec above 8°K $T_1 = 5\cdot5 \times 10^2 T^{-9}$ sec between 5° and 8°K	Possibly phonon bottleneck below 5°K	90
Er³⁺ in CaF₂ ~ 0·1%	C and I; 9 and 36; 1·6–25	$T_1 \propto T^{-4}$, 1·5°–4·2°K $T_1 \propto 10$ msec at 4·2°K	Relaxation time for main line much longer than for hyperfine lines	90
	C; 9·2 and 36; below 4·2	$T_1 \sim 5$ msec at 9·2 Gc/s $T_1 \sim 20$–40 msec at 36 Gc/s. Both at 2·25°K		91

	Band; frequency (Gc/s); temp. range (°K)	Results	Remarks	Ref.
Er³⁺ in CaF₂	B and C; X-band; up to 4·2	Several types of spectra referred to Spectrum (v) data given graphically Spectrum (VII) $T_1^{-1} = 9$ μsec at 4·2°K. 2·3–4·2°K, $T_1^{-1} = 0·34 \times 10^8$ exp$(-\Delta/kT)$ sec⁻¹, $\Delta = 18 \pm 2$ cm⁻¹ Below 2·3°K, $T_1 \propto T_1^{-1}$		92
1%	C; 9·6; 2–6	$H \parallel \langle 110 \rangle$. Relaxation studied on sites of tetragonal and trigonal symmetry Trigonal site: $T_1 = 9·6 \times 10^{-4} T^{-1} + 1·3 \times 10^{-12}$ exp$(78/T)$ sec Tetragonal site: $T_1 \approx 10^{-7}$ exp$(27/T)$ sec near 6°K; below 3°K, probably $T_1 \propto 1/T$		71
Er³⁺ in CaWO₄ 0·02%; paramagnetic ions substituted on Ca²⁺ sites	G and I; 9·4; 1·6–20	$T_1 \propto$ exp$(-\Delta/kT)$, where $\Delta/k \sim 30$°K below 4·2°K (pulse) and $\Delta/k \sim 70$°K above 4·2°K (linewidths) $T_1 \sim 10$ μsec at 4·2°K	Cross-relaxation effects serious at low temperatures. Line-broadening results not highly accurate. Ce³⁺ and Er³⁺ both present together. Primarily an investigation of phase memory times in spin echoes	56
Er³⁺ in CdF₂ ~0·1%	B and C; 9·2; 2–18	$T_1 = 2·75 \times 10^{-4}$ sec at 1·9°K $T_1 = 1·4 \times 10^{-4}$ sec at 4·2°K		93
0·03%	C; 3–70; 4·2	$T_1 = 400$ μsec at 38 Gc/s $T_1 = 200$ μsec at 71 Gc/s Frequency dependence observed $T_1^{-1} = \dfrac{5 \times 10^4}{\nu} + 0·86\nu^2$ sec⁻¹ (ν in Gc/s)	Extension of frequency range of reference 93, using same samples. For concentration of 2%, $T_1 \propto T^{-3·5}$	91
~0·1%	C and I; 9 and 36; 1·6–25	$T_1 \approx 10^{-3} T^{-1}$ sec below 6°K $T_1 = 2·5 \times 10^{-12}$ exp$(91/kT)$ sec above 8°K	Very little frequency variation observed below 4°K	90

A	B	C	D	E
Er³⁺ in La(C₂H₅SO₄)₃ · 9H₂O 0·2, 10 and 100%	C; X-band; 1·1–3·7	$T_1{}^{-1} = DT^2 + 1·6 \times 10^{11} \exp(-44 \times 1·44/T)$ sec⁻¹, where D varies from $2·5$–22 sec⁻¹ °K² over the concentration range	Angular dependence investigated	94
1%	C; 9·36; 1·2–4·6	$H \parallel z$-axis Experimental: $T_1{}^{-1} = 4·2T + 4·4 \times 10^{-3}T^9 + 4·5 \times 10^{10} \exp(-59/T)$ sec⁻¹ Theoretical: $T_1{}^{-1} = 0·15T + 1·1 \times 10^{-5}T^9 + 2·7 \times 10^9 \exp(-58/T)$ sec⁻¹ or $T_1{}^{-1} = 9·4T + 1·5 \times 10^{-3}T^9 + 7·1 \times 10^{10} \exp(-58/T)$ sec⁻¹	Measurements also performed for $H \perp z$-axis with similar temperature dependence. Slight angular dependence observed	85
Er³⁺ in LaCl₃ 1% (enriched to 97·9% even isotopes)	C; 8·89; 1·65–4·2	$H \perp c$-axis Experimental: $T_1{}^{-1} = 2·26T + 1·14 \times 10^{10} \exp(-54·6/T) + 3·46 \times 10^{-3}T^9$ sec⁻¹ Theoretical: $T_1{}^{-1} = 2·73T + 1·66 \times 10^{10} \exp(-54·6/T) + 0·679 \times 10^{-3}T^9$ sec⁻¹	Weak angular dependence observed	76
0·2–2%	C; 9·52; 1·1–10	$T_1{}^{-1} \approx 2·9T + 3·6 \times 10^9 \exp(-\Delta/kT) + 2·0 \times 10^{-3}T^9$ sec⁻¹, where Δ/k varies from 48·9°K to 54·5°K for different concentrations	Concentration dependence observed in Orbach term, but no angular variation	75
Er³⁺ in LaF₃ 0·05–1%	C; 9·3; 0·2–5	$H \parallel x$-axis Experimental: 0·05%; $T_1{}^{-1} = 0·35 \coth(0·22/T) + 1·8 \times 10^{-3}T^9 + 8·1 \times 10^{10} \exp(-72/T)$ sec⁻¹ 0·1–0·5%; $T_1{}^{-1} = AT + BT^9 + 7·5 \times 10^{10} \exp(-72/T)$ sec⁻¹ A and B depend non-uniformly on concentration Theoretical: $T_1{}^{-1} = 10T + 4·3 \times 10^{-4}T^9 + 1·2 \times 10^{11} \exp(-72/T)$ sec⁻¹	Concentration dependence attributed to cross relaxation to pairs or impurities. Results for the higher concentration samples could be fitted to a cosech(3·9/T) term appropriate to cross relaxation	77

Sample	Conditions	T_1 relations	Notes	Ref.
Er^{3+} in $La_3Al_5O_{12}$ Nominal 1%	C; 8·7; up to 20	$T_1^{-1} = 10T + 6 \times 10^7 \exp(-36 \times 1·44/T)$ sec⁻¹	First excited state at 36·7 cm⁻¹ (optical measurement)	87
Er^{3+} in $La_3Ga_5O_{12}$ Nominal 1%	C; 8·7; up to 20	$T_1^{-1} = 50T + 1·4 \times 10^9 \exp(-47 \times 1·44/T)$ sec⁻¹	First excited state at 49·7 cm⁻¹ (optical measurement)	87
Er^{3+} in SrF_2 0·1%	C and I; 36; 1·4-30	Trigonal sites $T_1^{-1} = 54 \coth(0·863/T) + 10^{-5}T^9 + 9 \times 10^{10} \exp(-122/T)$ sec⁻¹		89
Er^{3+} in $YCl_3 \cdot 6H_2O$ 0·1 and 1%	C; 9·36; 0·2-5	$H \parallel z$-axis, 1% sample $T_1^{-1} = 0·016 \coth(0·23/T) + 3·22 \times 10^6 \exp(-24·5/T)$ sec⁻¹ $H \perp z$, direct term $\sim 125T$	Angular dependence measured at 1·39°K for 0·1% sample	86
Er^{3+} in $Y(C_2H_5SO_4)_3 \cdot 9H_2O$ 1 and 0·1%	C; 9·31; 1·2-4·6	$H \perp z$-axis, 1% sample Experimental: $T_1^{-1} = 5·9T + 3 \times 10^{-3}T^9$ sec⁻¹ Theoretical: $T_1^{-1} = 0·26T + 3·5 \times 10^9 \exp(-63/T) + 7·6 \times 10^{-6}T^9$ sec⁻¹ or $4·5T + 9·2 \times 10^{10} \exp(-63/T) + 111 \times 10^{-3}T^9$ sec⁻¹	Experimental value of T_1 for 0·1% sample not significantly different	85
Er^{3+} in $Y_3Al_5O_{12}$ Nominal 1%	C; 8·7; up to 20	$T_1^{-1} = 20T + 3·2 \times 10^6 \exp(-20 \times 1·44/T)$ sec⁻¹	First excited state at 22·1 cm⁻¹ (optical measurement)	87
Er^{3+} in $Y_3Ga_5O_{12}$ Nominal 1%	C; 8·7; up to 20	$T_1^{-1} = 70T + 1·1 \times 10^9 \exp(-46 \times 1·44/T)$ sec⁻¹	First excited state at 44·3 cm⁻¹ (optical measurement)	87
En^{3+} in BaF_2 0·25%	B; 9·5; 4·2-90	$H \parallel (111)$ $T_1 = 90 \ \mu sec$ at 4·2°K $T_1 = 7·5 \ \mu sec$ at 90°K		95
$\sim 0·004\%$; grown by Stockbarger technique	B and G; 9; 1·4-13	$H \parallel (100)$ for $\pm\frac{1}{2}$ transition $T_1^{-1} = 33·3T + 2·79 \times 10^{-3}T^5$ sec⁻¹	Saturation measurements analysed by the method of Castner.[96] Iron present in similar concentration	53

A	B	C	D	E
Eu²⁺ in CaF₂ 0·0047, 0·0074 and 0·18%; spectrum complex	C; 8·9; 1·25–30	$H \parallel$ fourfold axis, $\pm\frac{1}{2}$ transition, pulse width $\sim$ 1 msec $T_1^{-1} = 12T + 5\cdot3 \times 10^{-4} T^5$ sec^{-1}	Fifth power temperature dependence observed over range 15–30°K characteristic of multilevel system (see Chapter 2). Pulse width variation investigated	97
Gd³⁺ in CaF₂ 0·03% by wt.; tetragonal site symmetry	C; 9·6; 2–45	$\pm\frac{1}{2}$ transition $H \parallel (100)$ $T_1 = 1\cdot1 \times 10^{-3} T^{-0\cdot5} + 4 \times 10^3 T^{-5}$ sec		71
Gd³⁺ in La(C₂H₅SO₄)₃ . 9H₂O Gd; La = 1; 200 Ce; La = 1; 500	B; 9·0; 1·2	$T_1 \sim 10^{-4}$ sec, varying with transition and angle. Cross relaxation to Ce impurity lines clearly seen		98
1%	B; 9; 4·2	$T_1 = 1\cdot6 \times 10^{-4}$ sec	Power at cavity 1×10^{-6} to 5×10^{-2} W. Absolute values probably only correct to order of magnitude. Derivatives normalized using sample of D.P.H.	16
5%	C; 8·75; 4·2	$H \parallel$ crystal axis $T_1 \sim 2\cdot6$–8·8 msec		23
Gd³⁺ in LaF₃ 0·1%	C; 9·3; 1·3–5	For $-\frac{7}{2}$ to $-\frac{5}{2}$ transition, $T_1^{-1} = 3120 \exp(-4\cdot45/T)$ sec^{-1}	Measurements made on all seven $\Delta m = 1$ transitions, for each of which different values of T_1 obtained. Recoveries usually non-exponential. Form of temperature dependence not satisfactorily explained; could be T^5 as expected for multilevel system	77

Gd^{3+} in La$_2$Mg$_3$(NO$_3$)$_{12}$. 24H$_2$O ~1%	B and modified C; 9.0 1-4	At 1.3°K on $\pm\frac{1}{2}$ transition T_1 (pulse) = 7×10^{-3} sec T_1 (saturation) = 15×10^{-3} sec	Saturation measurements probably unreliable due to broad background signal. Evidence of phonon bottleneck, lattice-bath relaxation probably measured	104
Gd^{3+} in SrS	C; 9.3; 1.6–4.2	$T_1 = 0.71T^{-1}$ sec		57
Gd^{3+} in ZrSiO$_4$ (zircon) 0.002%; natural crystal from Siam	B and I; X-band; 77, 290°	Single measurements $T_1 = 13\pm2$ μsec at 290°K $T_1 = 30\pm4$ μsec at 77°K	25 mW into cavity of $Q = 2500$ saturated the resonance at 290°K	99
Ho^{2+} in CaF$_2$ 0.02%	C; 8.9; 1.25–22	$H \parallel (100)$ Experimental: $T_1^{-1} = 42T + 8.0\times10^9$ $\exp\left(-\dfrac{33.1\times1.44}{T}\right)$ sec^{-1} Theoretical: $T_1^{-1} = 37T + 8.5\times10^9$ $\exp\left(-\dfrac{33.8\times1.44}{T}\right) + 8.5\times10^7$ $\exp\left(-\dfrac{30.1\times1.44}{T}\right) + 2.1\times10^{-4}\,T^9$ sec^{-1}		97
Ho^{3+} in LaCl$_3$	C; 9.52; 1.14	$T_1 = 3.5\pm1.0\times10^{-4}$ sec	Non-Kramers ion producing very weak signal	75
Nd^{3+} in BaMoO$_4$ 0.01%; grown by flux melt technique	C; 9.4; 1.5–70	$T_1^{-1} = 9.2T + 7.7\times10^{-4}T^9 + 6.6\times10^{11}\exp(-122/T)$ sec^{-1}		100

223

A	B	C	D	E
Nd^{3+} in CaF$_2$ 0.02–0.5%; both tetragonal and rhombic spectra investigated	C; 3.2–40; 1.5–2	$H \parallel$ tetragonal axis: $T_1 \sim 10$ to 100 msec at 1.5°K and 35 Gc/s, depending on concentration; Rhombic: $T_1 \sim 5$ msec at 1.8°K and 40 Gc/s, concentration dependent	Dependence on concentration, c, observed; $T_1 \propto 1/c^2$. $T_1 \propto H^{-2}$. Behaviour varied markedly between specimens	101
0.28% by wt.	C; 9.6; 2–7	$H \perp (100)$ $T_1 = 7.4 \times 10^{-4} T^{-1} + 2.2 \times 10^{-11} \exp(93/T)$ sec		71
Nd^{3+} in CaMoO$_4$ 0.5%; grown by Czochralski method	C; 9.4 and 36; 1.5–70	$T_1^{-1} = AT + 7 \times 10^{-6} T^9 + 5.1 \times 10^{11} \exp(-160/T)$ sec^{-1}, where $A = 8.3$ sec^{-1}°K^{-1} at 9.4 Gc/s, and $A = 23.8$ sec^{-1}°K^{-1} at 36 Gc/s	Sample grown by flux-melt method with $\sim 0.05\%$ Nd gave similar results at 9.4 Gc/s	100
Nd^{3+} in CaWO$_4$ 0.01–0.05%	C; 3.2–40; 1.5–2	$H \perp$ tetragonal axis $T_1 \sim 10$ msec at 40 Gc/s and 1.8°K Direct process dominates below 2°K	Magnetic field dependence investigated, $T_1 \propto H^{-4}$ at the higher frequencies	101
0.5%; grown by Czochralski method	C and I; 9.4 and 36; 1.5–70	$T_1^{-1} = AT + 4 \times 10^{-5} T^9 + 8.7 \times 10^{10} \exp(-136/T)$ sec^{-1}, where $A = 5$ sec^{-1}°K^{-1} at 9.4 Gc/s, and $A = 26$ sec^{-1}°K^{-1} at 36 Gc/s		100
	C; X-band; up to 15	$T_1 \sim 10$ msec at 4.2°K	Two phonon process above 6°K One phonon below and a form of bottleneck observed	102
0.008%	F; 9.4; 1.4–9	$H \parallel c$-axis $T_1^{-1} = 2.3T + BT^{10.4}$ sec^{-1}	Angular variation of direct process measured at 2.2°K	73

Nd³⁺ in CdMoO₄ 0·5%; grown by Czochralski method	C; 9·4; 1·5–70	$T_1^{-1} = 2 1T + 5\cdot8 \times 10^{-5}T^9$ $+ 1\cdot5 \times 10^{11}\exp(-154/T)\,\text{sec}^{-1}$		100
Nd³⁺ in LaCl₃ 0·25 and 2%	C; 9·52; 1·1–10	H at 38° to c-axis Experimental: $T_1^{-1} = 0\cdot53T + 3\cdot4 \times 10^{-5}T^9\,\text{sec}^{-1}$ for 0·25% sample $H \perp$ c-axis $T_1^{-1} = 2\cdot93T + 5\cdot3 \times 10^{-5}T^9$	Cross relaxation to Ce³⁺ ions detected. T_1 concentration dependent. Anisotropy in Raman process observed	75
0·1–2%	C; 9·0; 1·65–4·2	H at 45° to c-axis, 0·1% enriched sample Experimental: $T_1^{-1} = 0\cdot58T + 4\cdot85 \times 10^{-5}T^9\,\text{sec}^{-1}$ Theoretical: $T_1^{-1} = 0\cdot295T + 3\cdot75 \times 10^{-5}T^9\,\text{sec}^{-1}$	Possibly cross relaxation to Ce³⁺. Angular variation observed	76
Nd³⁺ in LaF₃ 0·1 and 1%	Modified C; 8·4–17·23; 0·18–3	$H \parallel z$-axis $T < 1\cdot5°$K, $T_1^{-1} \propto \coth \delta/2kT$ $T > 2°$K, $T_1^{-1} \propto \exp(6o/T)$ At c·2°K, $T_1 = 23$ sec For $T < 1\cdot5°$K, magnetic field dependence given by $T_1^{-1}\tanh(\delta/2kT) = 0\cdot86 \times 10^{-4}$ $c^2H + 2\cdot1 \times 10^{-12}H^3 + 4\cdot5 \times 10^{-19}H^5$ sec^{-1} (concentration c, field H)	H^5 term is probably Krönig–Van Vleck process. H^3 term could arise from h.f.s. relaxation processes. c^2H term possibly due to cross relaxation to pairs	103
0·1 and 1%	C; 9·23; 0·2–5	$H \parallel z$-axis, 0·1% Experimental: $T_1^{-1} = 0\cdot23T + 3\cdot6 \times 10^{10}\exp(-57/T)$ $+ 10^{-3}T^9\,\text{sec}^{-1}$ Theoretical: $T_1^{-1} = 0\cdot38T + 1\cdot1 \times 10^{10}\exp(-57/T)$ $+ 2\cdot7 \times 10^{-4}T^9\,\text{sec}^{-1}$ $H \parallel z$-axis, 1% Experimental: $T_1^{-1} = 0\cdot73T + 2\cdot4 \times 10^{11}\exp(-65/T)$ $+ 2\cdot4 \times 10^{-3}T^9\,\text{sec}^{-1}$	Concentration dependence of 'direct' term attributed to cross relaxation to pairs or clusters. 57°K for Δ/k gives a better fit to the experimental results, for the dilute sample, than the optically measured value of 65°K	77

A	B	C	D	E
Nd³⁺ in La(C₂H₅SO₄)₃·9H₂O 0·2% (enriched to 98·5% even isotopes), 0·5% and 5%	C; 9·37; 1·4–5	$H \perp z$-axis Experimental: $T_1^{-1} = 4\cdot4 T + 3\cdot65 \times 10^{-4} T^9$ sec⁻¹ (for 0·2% sample) Theoretical: $T_1^{-1} = 1\cdot4 T + 1\cdot3 \times 10^{-4} T^9$ sec⁻¹	'Direct' term increases with concentration, possibly due to an increasing amount of Ce³⁺ impurity. Cross-relaxation effects to Ce³⁺ ions investigated	74
Nd³⁺ in La₂Mg₃(NO₃)₁₂·24H₂O 1%, enriched to 98·5% even isotopes	Modified C; 8·4–17·23; 1·4–4·2	$H \perp$ crystal axis $T_1 = 8\cdot6$ msec at 2·61°K $T_1^{-1} = 9\cdot3 \times 10^9 \exp(-47\cdot6/T) + (T_1^*)^{-1}$ sec⁻¹, where $T_1^* = [1\cdot66\nu^4 \times 10^{-40}(\delta/2k)\coth(\delta/2kT)]^{-1} + [16(\delta/2k)^2\coth^2(\delta/2kT)]^{-1}$ sec	Orbach term independent of frequency. T_1 measured as function of frequency at 1·4°K where Orbach term negligible. coth² (bottleneck) term very frequency dependent. ν^4 dependence verified for 'direct' term	103
0·2%	F; 9·3; 1·5–4·2	$T_1 = 2 \times 10^{-9} \exp[(46\pm2)/T]$ $T_1 \sim 10^{-1}$ sec at 2·5°K		81
5% and 1% enriched to 98·5% even isotopes	C; 9·67; 0·3–2	$H \perp z$-axis Experimental: $T_1^{-1} = 6\cdot3 \times 10^9 \exp(-47\cdot6/T) + 0\cdot3\coth(h\nu/2kT)$ sec⁻¹ Theoretical: $T_1^{-1} = 2\cdot2 \times 10^{10} \exp(-47\cdot6/T) + 0\cdot6\coth(h\nu/2kT)$ sec⁻¹	Extension of the measurements of reference 74 to lower temperatures	84
5% natural abundance and 1% enriched to 98·5% even isotopes	C; 9·37 and 34·3; 1·4–5	Experimental: $T_1^{-1} = 1\cdot7T + 6\cdot3 \times 10^9 \exp(-47\cdot6/T)$ sec⁻¹ at 9·37 Gc/s $T_1^{-1} = 32T^2 + 4 \times 10^9 \exp(-47\cdot6/T)$ sec⁻¹ at 34·3 Gc/s Theoretical: $T_1^{-1} = 2\cdot6T + 2\cdot2 \times 10^{10} \exp(-47\cdot6/T) + 7\cdot8 \times 10^{-4} T^9$ sec⁻¹	No concentration dependence or cross-relaxation effects noticed. Incidence of bottleneck at higher frequencies consistent with theory	74

Nd³⁺ in Nd(C₂H₅SO₄)₃ · 9H₂O

100%	Modified C; 9·27; 1·6-4·2	$H \parallel z$-axis $T_1^{-1} = 333T$ sec^{-1}	Modified pulse-saturation method enables true spin-lattice relaxation time to be determined, even in the presence of a phonon bottleneck, by monitoring the reflected microwave pulse	105
100%	E; X-band; 1·4-4·2	$T_1 \sim 11$ msec at 4·2°K independent of H	Measurements made over range of magnetic fields 800-2600 gauss. T_1 strongly field dependent below 2°K, probably due to cross relaxation	106
100%	E; X-band; 1·4-4·2	$T_1 \sim 10$ msec at 4·2°K $T_1 \propto T^{-3}$ at all fields	T_1 decreases gradually with increasing magnetic field, with a broad maximum at 1500 gauss. Pulse lengths 8-124 msec	107

Nd³⁺ in PbMoO₄

0·5%	C; 9·4 and 36; 1·5-70	Experimental: $T_1^{-1} = AT + 2·2 \times 10^{-4}T^9$ $+ 4·4 \times 10^{11} \exp(-132/T)$ sec^{-1}, where $A = 0·74$ sec^{-1}°K^{-1} at 9·4 Gc/s and $A = 90$ sec^{-1}°K^{-1} at 36 Gc/s Theoretical: $T_1^{-1} = 0·043T + 1·7 \times 10^{-5}T^9$ $+ 2·4 \times 10^{11} \exp(-132/T)$ sec^{-1} at 9·4 Gc/s	Direct term very frequency dependent, $A \propto \nu^n$, where $n = 3·7 \pm 0·1$	100

Nd³⁺ in Y₃Al₅O₁₂

1%; rhombohedral field	C; 8·9; 1·25-20	$H \parallel (110)$ $T_1^{-1} = 34T$ $+ 4·5 \times 10^{10} \exp\left(-\frac{75 \times 1·44}{T}\right)$ sec^{-1}	No optical data for crystal field splittings to enable theoretical prediction to be made	97

Nd³⁺ in Y(C₂H₅SO₄)₃ · 9H₂O

0·1 and 1%	C; 9·4; 1·2-5	Measurements made on central line with $H \perp z$-axis Experimental: $T_1^{-1} = 1·2T + 1·64 \times 10^{-4}T^9$ sec^{-1} Theoretical: $T_1^{-1} = 0·24T + 3·3 \times 10^{-6}T^9$ or $0·74T + 3·4 \times 10^{-5}T^9$	No concentration dependence. Predicted value of T_1 depends on approximations made	85

227

A	B	C	D	E
Nd^{3+} in $YCl_3 \cdot 6H_2O$ 0·1 and 1 at. %	C; 9·36; 0·2–5	$H \parallel z$-axis $T_1^{-1} = 7 \times 10^{-3} \coth(0.23/T)$ $+ 4 \times 10^{11} \exp(-89/T) + 4.8 \times 10^{-47} T^9$ sec^{-1}	Large anisotropy in direct process observed at 1·45°K	86
Nd^{3+} in $Y_3Ga_5O_{12}$ Nominal 1%	C and I; 8·7; up to 20	$T_1^{-1} = 17T + 9 \times 10^{10} \exp(-85 \times 1.44/T)$	Some experimental data significantly outside equation given. T_1 values between 10^{-9} and 10^{-8} sec estimated from linewidths (15–20°K)	87
1% rhombohedral	C; and I 8·9; 1·25–20	$H \parallel (110)$ $T_1^{-1} = 17T + 9.0 \times 10^{10} \exp\left(-\dfrac{85 \times 1.44}{T}\right)$ sec^{-1}		97
Pr^{3+} in $LaCl_3$ 0·3 and 1%	C; 9·17; 1·65–4·2	$H \parallel c$-axis, 1% sample $T_1^{-1} = 272T^2 + 2.06 \times 10^9 \exp(-47.9/T)$ sec^{-1}	Angular and concentration dependence of bottleneck term investigated	76
Pr^{3+} in $La(C_2H_5SO_4)_3 \cdot 9H_2O$ 5%	C; 9·27; 1·2–3	$H \parallel z$-axis. $M_I = \frac{3}{2}$ transition Experimental: $T_1^{-1} = 5 \times 10^3 T^2 + 6.5 \times 10^7 \exp(-21/T)$ sec^{-1} $T_1 \sim 50$ μsec at 2°K (direct process estimated as $> 5 \times 10^4 T$ from $AT \gg DT^2$) Theoretical: $T_1^{-1} = 3.6 \times 10^4 T + 5.9 \times 10^7 \exp(-20/T) + 6.2 \times 10^{-3} T$ sec^{-1} or $5 \times 10^5 T + 4.2 \times 10^9 \exp(-20/T) + 1.7 T^7$ sec^{-1} depending on approximations	Crystal heating effect noticed at high microwave pump powers as discontinuity below λ point. $T < 2.17°K$, $T_1 \sim 35$ μsec. Above λ point, $T > 2.17°K$, $T_1 \sim 37$ msec	85
Pr^{3+} in $La_2Mg_3(NO_3)_{12} \cdot 24H_2O$ 0·6%	F; 9·3; 1·5–4·2	$T_1 = 0.36T^{-6.5 \pm 0.5}$ sec $T_1 \sim 10^{-3}$ sec at 2·5°K		81

228

Pr³⁺ in La₂Mg₃(NO₃)₁₂ · 24H₂O 0·1 and 1%	C; 9·15; 1·4–4·2	Experimental: $T_1^{-1} = 84T+4\cdot6$ $\times 10^{10}\exp(-54\cdot6/T)+2\cdot35T^7$ sec^{-1} (1% sample) $T_1^{-1} = 500T^2+2\cdot35T^7$ (0·1% sample) Theoretical: $T_1^{-1} = 7\cdot4\times10^4T+9\cdot7$ $\times10^9\exp(-54\cdot6/T)+0\cdot14T^7$ sec^{-1}	Non-Kramers ion	74.
Pr³⁺ in Pr(C₂H₅SO₄)₃ · 9H₂O 100%	E; X-band; 1·4–2·18 and 4·2	$T_1 \sim 0\cdot2$ msec at 1·8°K	Evidence that this is phonon-bath relaxation time	108
P³⁺ in Y(C₂H₅SO₄)₃ · 9H₂O 5%	C; 9·4; 1·2–3	$H \parallel z$-axis Experimental: $T_1^{-1} = 3\cdot8\times10^7\exp(-19/T)$ $+4\cdot1\times10^3T^2$ sec^{-1} Theoretical: $T_1^{-1} = 3\cdot6\times10^4T$ $+5\cdot9\times10^7\exp(-20/T)+6\cdot2\times10^{-3}T^7$ sec^{-1} or $T_1^{-1} = 5\times10^5T+4\cdot2\times10^9\exp(-20/T)$ $+1\cdot7T^7$ sec^{-1} depending on approximations		85
Sm³⁺ in LaCl₃ 0·1%	C; 9·15; 1·65–4·2	$H \parallel c$-axis $T_1^{-1} = AT+3\cdot4\times10^{11}\exp(-58\cdot6/T)$ $+0\cdot02T^9$ sec^{-1}, where A varies from 0·035– 3·34 sec^{-1}°K^{-1} for 3 specimens A(calculated) $= 0\cdot47$ sec^{-1}°K^{-1}	Difficulty experienced in growing Sm³⁺ doped specimens. Spectral diffusion effects observed. Angular variation investigated	76
	0·2 and 0·5%	C; 9·52; 1–4·2	$H \parallel c$-axis $T_1^{-1} = 0\cdot9T+0\cdot016T^9$ $+3\cdot65\times10^9\exp(-44\cdot3/T)$ sec^{-1}	Angular variation noted for 'direct' term	75
Sm³⁺ in La(C₂H₅SO₄)₃ · 9H₂O 0·1%	C; 8·9; 1·25–5	Experimental: $T_1^{-1} = AT+2\cdot6\times10^{-2}T^9$ sec^{-1}, where $A = 16$ for $H\parallel c$-axis, $A = 100$ sec^{-1}°K^{-1} for $H\perp c$-axis Theoretical: $T_1^{-1} = AT+44\times10^{-3}T^9$ sec^{-1}, where $A = 0\cdot32$ sec^{-1}°K^{-1} for $H\parallel c$-axis, $A = 1\cdot8$ sec^{-1}°K^{-1} for $H\perp c$-axis	This is an experimental observation of the anisotropy predicted for the direct process by Orbach	97

A	B	C	D	E
Sm^{3+} in La(C$_2$H$_5$SO$_4$)$_3$ · 9H$_2$O 0·1 and 1%	C; 9·4; 14–5	$H \parallel z$-axis Experimental: $T_1^{-1} = 1\cdot0T + 3\cdot1 \times 10^{-3}T^9$ $+ 5\cdot8 \times 10^8 \exp(-46/T)$ $+ 6\cdot1 \times 10^{10} \exp(-72/T)$ sec^{-1} Theoretical: $T_1^{-1} = 0\cdot69T + 0\cdot89 \times 10^{-5}T^9$ $+ 1\cdot5 \times 10^9 \exp(-46/T)$ $+ 3\cdot1 \times 10^9 \exp(-72/T)$ sec^{-1} or $6\cdot1T$ $+ 3\cdot7 \times 10^{-4}T^9 + 1\cdot8 \times 10^{10} \exp(-46/T)$ $+ 5\cdot4 \times 10^{10} \exp(-72/T)$ sec^{-1}	Evidence of a second Orbach process via an unexpectedly high intermediate level. Long pulse greater than 1 msec used to reduce cross-relaxation effects. Theory and experiment also compared for $H \perp z$-axis	85
Sm^{3+} in LaF$_3$ 0·5%	C; 9·3; 14–5	$H \parallel c$-axis Experimental: $T_1^{-1} = 0\cdot43T^2$ $+ 1\cdot5 \times 10^{10} \exp(-50/T)$ sec^{-1} $H \parallel y$-axis Experimental: $T_1^{-1} = 9\cdot2T$ $+ 1\cdot5 \times 10^{10} \exp(-50/T)$ sec^{-1}	T^2 dependence attributed to cross-relaxation not bottleneck	77
Sm^{3+} in La$_2$Mg$_3$(NO$_3$)$_{12}$ · 24H$_2$O ~ 1% from line intensity measurements	C; 9·38; 1·2–4	$H \parallel z$-axis Experimental: $T_1^{-1} = 3\cdot4T + 1\cdot3 \times 10^{-2}T^9$ $+ 1\cdot6 \times 10^{10} \exp(-55/T)$ sec^{-1} Theoretical: $T_1^{-1} = 3\cdot4T + 1\cdot5 \times 10^{-3}T^9$ $+ 4\cdot4 \times 10^{10} \exp(-55/T)$ sec^{-1}	Experimental value of Δ/k accurate to $\pm 3°$K	85
0·05%	C; 9·25; 1·4–2·2	$H \parallel z$-axis Experimental: $T_1^{-1} = 8T + 4 \times 10^{-3}T^9$ sec^{-1} Theoretical: $T_1^{-1} = 80T + 2\cdot5 \times 10^{-2}T^9$ sec^{-1}	Zero-order calculation not sufficient to account for relaxation behaviour; perturbation from higher level must be considered	74
Sm^{3+} in Sm(C$_2$H$_5$SO$_4$)$_3$ · 9H$_2$O 100%	C; 9·4; 1·5–5	$H \parallel z$-axis Experimental: $T_1^{-1} = 9\cdot5 \times 10^8 \exp(-51/T)$ $+ 5\cdot8 \times 10^{-4}T^9$ sec^{-1}		85

230

Sm³⁺ in Sm₂Mg₃(NO₃)₁₂·24H₂O 100% C;9·34;1·4–3	$H \perp z$-axis Experimental: $T_1^{-1} = 1·32T^2 + 5 \times 10^{-2}T^9$ sec⁻¹ Theoretical: $T_1^{-1} = 120T + 4·6 \times 10^{-2}T^9$ sec⁻¹	Phonon bottleneck expected on theoretical grounds	74
Sm³⁺ in Y(C₂H₅SO₄)₃·9H₂O 1% C;9·4;1·5–5	$H \parallel z$-axis Experimental: $T_1^{-1} = 0·76T + 4 \times 10^{-4}T^9$ $+ 8 \times 10^8 \exp(-51/T)$ sec⁻¹ Theoretical: $T_1^{-1} = 0·56T$ $+ 3·7 \times 10^9 \exp(-51/T) + 1·3 \times 10^{-6}T^9$ sec⁻¹ or $5·0T + 2·4 \times 10^{10} \exp(-51/T)$ $+ 8·4 \times 10^{-4}T^9$ sec⁻¹	Experimental value of Δ/k accurate to 2°K. Measured 'direct' term $= 1·3T$ for $H \perp z$-axis	85
Tb³⁺ in CaF₂ C;9·6;2–10	$H \perp (111)$ $T_1 \propto T^{-1}$ below 4°K $T_1 \sim 20$ μsec at 3°K	Angular variation of T_1 observed	71
Tb³⁺ in CaWO₄ 0·03 at.%; grown by Czochralski technique, charge compensation by Na C;9·4;1·3–9	Experimental: $T_1^{-1} = 2·67T + 7·7 \times 10^{-4}T^7$ $- 6·67 \times 10^8 \exp(-78·8/T)$ sec⁻¹ Theoretical: $T_1^{-1} = 1·4T + 2·6 \times 10^{-3}T^7$ $+ 6 \times 10^7 \exp(-78·8/T)$ sec⁻¹ $H \parallel c$-axis only. Measurements on high-frequency line from nuclear level $m_I = +\frac{1}{2}$	Pulse length increased until exponential recovery observed. Monitor power reduced until further reduction did not affect recovery time constant	109
Tb³⁺ in LaCl₃ 0·1 and 1% C;8·85;1·65–4·2	$H \parallel c$-axis Experimental: $T_1^{-1} = 4·26T + 4·83$ $\times 10^{11} \exp(-81·8/T) + 0·027T^7$ sec⁻¹ Theoretical: $T_1^{-1} = 1·01T + 1·18$ $\times 10^{11} \exp(-81·8/T) + 0·065T^7$ sec⁻¹	Angular dependence observed and predicted for 'direct' term	76

A	B	C	D	E
Tb^{3+} in Tb(C$_2$H$_5$SO$_4$)$_3$ · 9H$_2$O 100%	Modified E; —; 1·4–3·5	$T_1 \sim 6$–10 msec at 2·13°K	Magnetic field pulse used to disturb populations. Measured value of T_1 depends on magnitude of pulse. Phonon bottleneck unambiguously detected	110, 111
Tb^{3+} in Y(C$_2$H$_5$SO$_4$)$_3$ · 9H$_2$O 1%	C; 12–17; 1·5–5	$H \parallel z$-axis, 17 Gc/s, $M_I = \frac{1}{2}$ line; Experimental: $T_1^{-1} = 59T + 0·92 \times 10^{-2} T^7$ sec^{-1}; Theoretical: $T_1^{-1} = 67·4T + 4·8 \times 10^{-3} T^7$ sec^{-1}	Measurements also performed in zero magnetic field at 12·0 and 14·93 Gc/s. Similar temperature variations found; frequency dependence not correctly given	85
Tm^{2+} in CaF$_2$	C; 8·9; 1·25–22	$H \parallel (100)$; Experimental: $T_1^{-1} = 13T + 7·7 \times 10^{-8} T^9$ sec^{-1}; Theoretical: $T_1^{-1} = 4·5T + 1·9 \times 10^{-6} T^9$ sec^{-1}		97
Yb^{3+} in CaF$_2$ 0·17%; cubic and tetragonal sites studied	C; 9·6; 2–20	Tetragonal sites; For both $g_{\parallel}$ and $g_{\perp}$ resonances, $T_1^{-1} = 1·2 \times 10^{-3} T^{-1} + 10^{-4} T^{-9}$ [last term also expressible as exp(125/T)]; Cubic sites; Low temperature, $T_1 \propto T^{-1}$, $5 < T < 20°$K, $T_1 \propto T^{-1} + \beta T(-3·5 \pm 0·5)$ [or $+ \beta' \exp(38/T)$]; $T_1 \sim 300$ μsec at 4°K	Similar measured values on other samples. Detailed discussion of result	71
Yb^{3+} in CaWO$_4$ 0·004%	F; 9·4; 1·4–9	$H \parallel c$-axis; $T_1^{-1} = 8·1T + BT^{10·1}$ sec^{-1}	Angular variation for T_1 even in Raman region (at 7°K)	73
Yb^{3+} in CdF$_2$ 0·1% by weight	B and C; 9·1; 1·8 and 4·2	$T_1 = (6·7 \pm 0·7) \times 10^{-4} T^{-1}$ sec		112

232

Material	C and I; frequency (Gc/s); temp (°K)	Relaxation data	Comment	Ref
Yb^{3+} in CdF$_2$ 0·01 to 0·1%	C and I; 3·2 and 9·1; 2-77	Below 4°K, $T_1 \propto T^{-1}$, concentration dependent T_1 also approx. proportional to frequency $T_1 \sim 3 \times 10^{-4}$ sec at 2°K $T_1 \sim 57 \times 10^{-9}$ sec at 57°K		113
0·01, 0·05, 0·1% in melt	C and I; 3·2 and 9·1; 2-77	Below 4·2°K, $T_1 \propto T^{-1}$ and (concentration)$^{-1}$ 20-77°K, $T_1 \propto T^{-7}$ At 4·2°K, T_1 (9·1 Gc/s) is approximately three times T_1 (3·2 Gc/s) $T_1 \sim 1$ msec at 4·2°K and 9·1 Gc/s		114
Yb^{3+} in LaF$_3$ 1%	C; 9·4; 0·2-5	$H \parallel z$-axis Experimental: $T_1^{-1} = 6·95[1+ \exp(0·47/T)]^{-1}+111 \operatorname{cosech}(3·9/T) +3 \times 10^{-3}T^9$ sec^{-1} $H \parallel c$-axis Experimental: $T_1^{-1} = 1790 \operatorname{cosech}(5·7/T) +3 \times 10^{-3}T^9$ sec^{-1}	Estimate of direct relaxation rate is $< 3T$ sec^{-1}. Cosech terms indicate cross relaxation to pairs	77
Yb^{3+} in Lu$_3$Ga$_5$O$_{12}$ 1%	C; 8·9; 1·25-20	$H \parallel (110)$ Experimental: $T_1^{-1} = 9·8T^{1·7}+1·0 \times 10^{-7}T^9$ sec^{-1} Theoretical: $T_1^{-1} = 3·1T+3·2 \times 10^{-7}T^9$ sec^{-1}	$T^{1·7}$ term attributed to presence of pairs. A direct term fitted to 1·3°K would contribute 12T to relaxation rate	97
Yb^{3+} in Y$_3$Al$_5$O$_{12}$ 0·1 and 1%	C; 8·9; 1·25-20	Experimental: $T_1^{-1} = 15T+6·3 \times 10^{-7}T^9$ sec^{-1} Theoretical: $T_1^{-1} = 5·3T+9·0 \times 10^{-7}T^9$ sec^{-1}	Concentration dependence attributed to contributions from pairs	97
Yb^{3} in Y$_3$Ga$_5$O$_{12}$ 0·1-10%	C; 8·9; 1·25-20	$H \parallel (100)$ for 0·1% sample Experimental: $T_1^{-1} = 33T+1·8 \times 10^{-7}T^9$ sec^{-1} Theoretical: $T_1^{-1} = 4·2T+5·7 \times 10^{-7}T^9$ sec^{-1}	Concentration dependence attributed to contribution from pairs	97

A	B	C	D	E
Yb^{3+} in Y$_3$Ga$_5$O$_{12}$ 0·1 and 1%	C; 8·7; up to 20, 77	0·1% sample: $T_1^{-1} = 33T + 1\cdot8 \times 10^{-7}T^9$ sec^{-1}, below 20°K. At 77°K, $T_1 \sim 8 \times 10^{-10}$ sec, 20 times greater than extrapolated value 1% sample: Similar results 10–20°K. Below 10°K, $T_1^{-1} = 30T^{1\cdot4}$ approximately	'Crude' theoretical estimate of direct term 10^2 to 10^3 longer than observed	87

234

A	B	C	D	E
Ir^{4+} in $(NH_4)_2PtCl_6$ 0·19–8·5%	C; 9·5; 1·2–5	Measurements on single ions, and pairs Single ions: $1/T_1 \propto AT + BT^9 + Cc^2 \times \exp(-5\cdot5/T)$, where c is the concentration $1/T_1$ varied from 3·0 to 1 × 10^4 sec^{-1} at 2°K, with concentration Pairs: Low concentrations: $1/T_1 \propto [1 - \exp(-\mathcal{J}/kT)]^{-1}$, concentration dependent	Cross relaxation to triads used to explain single ion results	117
Ir^{4+} in K_2PtCl_6 0·2–8·5%	C; 9·5 and 23·3; 1·4–4·2	$T_1^{-1} = A'f^4T + BT^9 + 1/T_{1c}$, where last term is dependent on concentration c $A'f^4 = 1\cdot9$ sec^{-1} degK^{-1} at 9·5 Gc/s $B = 3\cdot5 \times 10^{-2}$ sec^{-1} degK^{-9} $1/T_{1c}$ has form $f(c) \exp(-\Delta/kT)$, where $f(c) = c^2$ at low concentrations	Detailed theoretical treatment	118
Ir^{4+} in $(NH_4)_2PtCl_6$ 0·2–8·5%	C; 9·5 and 23·3; 1·4–4·2	$T_1^{-1} = A'f^4T + BT^9 + 1/T_{1c}$, where the last term is dependent on concentration c $A' = 2\cdot1 \times 10^{-40}$ sec^{-5} degK^{-1} $B = 5 \times 10^{-5}$ sec^{-1} degK^{-9} $1/T_{1c}$ has form $f(c) \exp(-\Delta/RT)$, where $f(c) = c^2$ at low concentrations	Detailed theoretical treatment	118
Pt^{3+} in Al_2O_3	I; 9·3; 77–500	$T_1 = (4\pm1) \times 10^{-11} \exp[(470 \pm 40)/kT]$ sec	Jahn–Teller splitting of ^2E ground state	66

235

A	B	C	D	E
Pa^{4+} in Cs$_2$ZrCl$_6$ 1 at. %	C; 9·1; 1·4–4·2	$T_1^{-1} = 3\cdot46T + 5\cdot1 \times 10^{-5}T^9$ sec^{-1}	Signal strength temporarily enhanced by annealing. Pulse duration 1 msec. No variation in T_1 amongst hyperfine lines. Coupling coefficients estimated theoretically using Orbach's approach are in reasonable agreement with those calculated from the experimental relaxation parameters	115
Rh^{2+} in ZnWO$_4$ 0·0045 and 0·06 at. %	C, B and I; 9·2; 1·5–110	56–110°K, $T_1 = 1\cdot7 \times 10^{-11} \exp(460/T)$ sec, independent of orientation and Rh concentration 1·5–4·2°K, Angular variation in T_1 Temperature variation T^{-3} to $T^{-4\cdot9}$ $T_1 \sim 1$ msec at 4·2°K	No variation in T_1 found with pulse length, over range 5 μsec to 10 msec. Theoretical discussion	116

References

1. Zverev, G. M., and Petelina, N. G., *Sov. Phys. JETP*, 1962, **15**, 820.
2. Taylor, A. G., Olsen, L. C., Brice, D. K., and Culvahouse, J. W., *Phys. Rev.*, 1966, **152**, 403.
3. Pryce, M. H. L., *Proc. Roy. Soc.*, 1965, **A283**, 433.
4. Atsartin, V. A., Morshnev, S. K., and Potkin, L. I., *Sov. Phys. Solid St.*, 1967, **9**, 660.
5. Zverev, G. M., and Prokhorov, A. M., *Sov. Phys. JETP*, 1963, **16**, 303.
6. Standley, K. J., and Tooke, A. O., *J. Phys. C.*, 1968, **1**, 149.
7. Nisida, Y., *J. Phys. Soc. Japan*, 1964, **19**, 2273.
8. Eschenfelder, A. H., and Weidner, R. T., *Phys. Rev.*, 1953, **92**, 869.
9. Squire, P. T., and Orton, J. W., *Proc. Phys. Soc.*, 1966, **88**, 649.
10. Andreeva, E. B., Karlov, N. V., Manenkov, A. A., Milyaev, V. A., and Shirkov, A. V., *Sov. Phys. Solid St.*, 1964, **6**, 1293.
11. Turoff, R. D., *Phys. Rev.*, 1965, **138**, 1524.
12. Kissinger, P. B., and Weidner, R. T., *Phys. Rev.*, 1962, **126**, 506.
13. Castle, J. G., Chester, P. F., and Wagner, P. E., *Phys. Rev.*, 1960, **119**, 953.
14. Kaplan, D. E., Browne, M. E., and Cowen, J. A., *Rev. Sci. Inst.*, 1961, **32**, 1182.
15. Kipling, A. L., Smith, P. W., Vanier, J., and Woonton, G. A., *Canad. J. Phys.*, 1961, **39**, 1859.
16. Marr, G. V., and Swarup, P., *Canad. J. Phys.*, 1960, **38**, 495.
17. Pace, J. H., Sampson, D. F., and Thorp, J. S., *R.R.E. Memorandum*, 1752 (1960).
18. Pashinin, P. P., and Prokhorov, A. M., *Sov. Phys. JETP*, 1961, **13**, 33.
19. Tengblad, R. G., and Yngvesson, K. S., *Phys. Letters*, 1967, **25A**, 437.
20. Castle, J. G., and Feldman, D. W., *Phys. Rev.*, 1961, **121**, 1349.
21. Atsartin, V. A., and Popov, V. I., *Sov. Phys. JETP*, 1965, **20**, 578.
22. Atsartin, V. A., Gerasimova, E. A., Matereeva, I. G., and Frantsesson, A. V., *Sov. Phys. JETP*, 1963, **16**, 903.
23. Davies, C. F., Strandberg, M. W. P., and Khyl, R. L., *Phys. Rev.*, 1958, **111**, 1268.
24. Pace, J. H., Sampson, D. F., and Thorp, J. S., *Proc. Phys. Soc.*, 1961, **77**, 257.
25. Lampe, D. R., and Wagner, P. E., *J. Chem. Phys.*, 1966, **45**, 1405.
26. Manenkov, A. A., Milyaev, V. A., and Prokhorov, A. M., *Sov. Phys. Solid St.*, 1962, **4**, 280.
27. Atsartin, V. A., Litovkina, L. P., and Meil'man, M. L., *Sov. Phys. Solid St.*, 1966, **7**, 2502.
28. Orton, J. W., Fruin, A. S., and Walling, J. C., *Proc. Phys. Soc.*, 1966, **87**, 703.
29. Emel'yanova, E. N., Karlov, N. V., Manenkov, A. A., Milyaev, V. A., Prokhorov, A. M., Smirnov, S. P., and Shirkov, A. V., *Sov. Phys. JETP*, 1963, **16**, 903.
30. Standley, K. J., and Wright, J. K., *Proc. Phys. Soc.*, 1964, **83**, 361.
31. Taylor, P. F., Ph.D. Thesis, University of St Andrews, 1968.
32. Breen, D. P., Krupka, D. C., and Williams, F. I. B., *Phys. Rev.*, to be published.
33. Lee, K. P., and Walsh, D., *Phys. Letters*, 1968, **27A**, 17.
34. Nash, F. R., *Phys. Rev. Letters*, 1961, **7**, 59.
35. Gill, J. C., *Proc. Phys. Soc.*, 1965, **85**, 119.
36. Stoneham, M., *Proc. Phys. Soc.*, 1965, **85**, 107.
37. Nash, F. R., *Phys. Rev.*, 1965, **138**, 1500.
38. Riggs, R. J., Ph.D. Thesis, University of St Andrews, 1968.
39. Lewis, M. F., and Stoneham, A. M., *Phys. Rev.*, 1967, **164**, 271.
40. Jones, J. B., and Lewis, M. F., *Solid St. Commun.*, 1967, **5**, 595.
41. Lewis, M. F., *Solid St. Commun.*, 1967, **5**, 845.
42. Kask, N. E., Kornienko, L. S., and Smirnov, A. I., *Sov. Phys. Solid St.*, 1963, **5**, 1212.

43. Madan, M. P., *Canad. J. Phys.*, 1964, **42**, 583.
44. Bray, T., Brown, G. C., Jr., and Kiel, A., *Phys. Rev.*, 1962, **127**, 730.
45. David, R. F., and Wagner, P. E., *Phys. Rev.*, 1966, **150**, 192.
46. Weissfloch, C. F., *Canad. J. Phys.*, 1966, **44**, 3185.
47. Paxman, D. H., *Proc. Phys. Soc.*, 1961, **78**, 180.
48. Pashinin, P. P., and Prokhorov, A. M., *Sov. Phys. Solid St.*, 1964, **5**, 1990.
49. Prokhorov, A. M., and Fedorov, V. B., *Sov. Phys. JETP*, 1964, **19**, 1305.
50. Rannestad, A., and Wagner, P. E., *Phys. Rev.*, 1963, **131**, 1953.
51. Carter, D. L., and Okaya, A., *Phys. Rev.*, 1960, **118**, 1485.
52. Orton, J. W., Private communication.
53. Horak, J. B., and Nolle, A. W., *Phys. Rev.*, 1967, **153**, 372.
54. Min-tsong, Lay F., and Nolle, A. W., *Phys. Rev.*, 1967, **163**, 266.
55. Sumita, M., *Jap. J. Appl. Phys.*, 1967, **6**, 1027.
56. Mims, W. B., *Phys. Rev.*, 1968, **168**, 370.
57. Manenkov, A. A., and Milyaev, V. A., *Paramagnetic Resonance*, Vol. II (Academic Press, New York, 1963), p. 419.
58. Fanton, J. C., and Averbuck, P., *Electronic Magnetic Resonance and Solid Dielectrics*, 12th Colloque Ampère (North Holland, Amsterdam, 1964), p. 354.
59. Soloman, P. R., *Phys. Rev.*, 1966, **152**, 452.
60. Turoff, R. D., Coulter, R., Irish, J., Sundquist, M., and Buchner, E., *Phys. Rev.*, 1967, **164**, 406.
61. Persyn, G. A., and Nolle, A. W., *Phys. Rev.*, 1965, **140**, A1610.
62. Manenkov, A. A., and Milyaev, V. A., *Sov. Phys. JETP*, 1962, **14**, 75.
63. Imbusch, G. F., Chinn, S. R., and Geschwind, S. G., *Phys. Rev.*, 1967, **161**, 295.
64. Bowers, K. D., and Mims, W. B., *Phys. Rev.*, 1959, **115**, 285.
65. Valishev, R. M., *Sov. Phys. Solid St.*, 1965, **7**, 733.
66. Höchli, V., and Muller, K. A., *Phys. Rev. Letters*, 1964, **12**, 730.
67. Kask, N. E., Kornienko, L. S., Mandel'shtam, T. S., and Prokhorov, A. M., *Sov. Phys. Solid St.*, 1964, **5**, 1677.
68. Kornienko, L. S., and Prokhorov, A. M., *Sov. Phys. JETP*, 1960, **11**, 1189.
69. Sanders, R. L., and Rowan, L. G., *Phys. Rev. Letters*, 1968, **21**, 140.
70. Zverev, G. M., *Sov. Phys. JETP*, 1963, **17**, 1251.
71. Bierig, R. W., Weber, M. J., and Warshaw, S. I., *Phys. Rev.*, 1964, **134**, A1504.
72. Vaughan, R. A., Ph.D. Thesis, University of Nottingham, 1964.
73. Kiel, A., and Mims, W. B., *Phys. Rev.*, 1967, **161**, 386.
74. Scott, P. L., and Jeffries, C. D., *Phys. Rev.*, 1962, **127**, 32.
75. Mangum, B. W., and Hudson, R. P., *J. Chem. Phys.*, 1966, **44**, 704.
76. Mikkelson, R. C., and Stapleton, H. J., *Phys. Rev.*, 1965, **140**, A1968.
77. Schulz, M. B., and Jeffries, C. D., *Phys. Rev.*, 1966, **149**, 270.
78. Brya, W. J., and Wagner, P. E., *Phys. Rev.*, 1966, **147**, 239.
79. Brya, W. J., and Wagner, P. E., *Phys. Rev.*, 1967, **157**, 400.
80. Cowen, J. A., and Kaplan, D. E., *Phys. Rev.*, 1961, **124**, 1098.
81. Cowen, J. A., Kaplan, D. E., and Browne, M. E., *J. Phys. Soc., Japan*, 1962, **17**, Suppl. B1., p. 465.
82. Kaplan, D. E., Browne, M. E., and Cowen, J. A., *Rev. Sci. Inst.*, 1961, **32**, 1182.
83. Leifson, O. S., and Jeffries, C. D., *Phys. Rev.*, 1961, **122**, 1781.
84. Ruby, R. H., Benoit, H., and Jeffries, C. D., *Phys. Rev.*, 1962, **127**, 52.
85. Larson, G. H., and Jeffries, C. D., *Phys. Rev.*, 1966, **141**, 461.
86. Schulz, M. B., and Jeffries, C. D., *Phys. Rev.*, 1967, **159**, 277.
87. Svare, I., and Seidel, G., *Paramagnetic Resonance*, Vol. II (Academic Press, New York, 1963), p. 430.
88. Gill, J. C., *Proc. Phys. Soc.*, 1963, **82**, 1066.
89. Antipin, A. A., Katyshev, A. N., Kurkin, I. N., and Shekun, L. Ya., *Sov. Phys. Solid St.*, 1967, **9**, 1070.
90. Zverev, G. M., and Smirnov, A. I., *Sov. Phys. Solid St.*, 1964, **6**, 76.
91. Zverev, G. M., and Smirnov, A. I., *Sov. Phys. Solid St.*, 1966, **8**, 1101.

92. Bobrovnikov, Y. U., Zverev, G. M., and Smirnov, A. I., *Sov. Phys. Solid St.*, 1967, **8**, 1750.
93. Zverev, G. M., Kornienko, L. S., Prokhorov, A. M., and Smirnov, A. I., *Sov. Phys. Solid St.*, 1962, **4**, 284.
94. Viswanathan, C. R., and Kaelin, G., *Phys. Rev.*, 1968, **171**, 992.
95. Miller, J. R., and Makendroo, P. P., *Phys. Rev.*, 1968, **174**, 369.
96. Castner, T. G., *Phys. Rev.*, 1959, **115**, 1058.
97. Huang, Chao-Yuan, *Phys. Rev.*, 1965, **139**, A241.
98. Feher, G., and Scovil, H. E. D., *Phys. Rev.*, 1957, **105**, 760.
99. Halton, D. R., and Troup, G. J., *Brit. J. Appl. Phys.*, 1964, **15**, 405.
100. Antipin, A. A., Katyshev, N., Kurkin, I. N., and Shekun, L. Ya., *Sov. Phys. Solid St.*, 1967, **9**, 636.
101. Kask, N. E., *Sov. Phys. Solid St.*, 1966, **8**, 900.
102. Kask, N. E., Kornienko, L. S., Prokhorov, A. M., and Fakir, M., *Sov. Phys. Solid St.*, 1964, **5**, 1675.
103. Baker, J. M., and Ford, N. C., *Phys. Rev.*, 1964, **136**, A1692.
104. Giordmaine, J. A., Alsop, L. E., Nash, F. R., and Townes, C. H., *Phys. Rev.*, 1958, **109**, 302.
105. Pfeuffer, K., *Phys. Stat. Sol.*, 1968, **29**, 171.
106. Rieckhoff, K. E., and Griffiths, D. J., *Canad. J. Phys.*, 1963, **41**, 33.
107. Daniels, J. M., and Rieckhoff, K. E., *Canad. J. Phys.*, 1960, **38**, 604.
108. Griffiths, D. J., and Glättli, H., *Canad. J. Phys.*, 1965, **43**, 2361.
109. Breen, D. B., and Coupland, G. M., *J. Phys. C.*, 1968, **1**, 146.
110. Kahle, H. G., Kalbfleisch, H., and Kump, U., *Z. Phys.*, 1965, **188**, 193.
111. Kahle, H. G., Kalbfleisch, H., and Stein, E., *Phys. Stat. Sol.*, 1967, **22**, 537.
112. Konyukov, V. K., Pashinin, P. P., and Prokhorov, A. M., *Sov. Phys. Solid St.*, 1962, **4**, 175.
113. Pashinin, P. P., and Prokhorov, A. M., *Sov. Phys. Solid St.*, 1963, **5**, 261.
114. Pashinin, P. P., and Prokhorov, A. M., *Paramagnetic Resonance*, Vol. 1 (Academic Press, New York, 1963), p. 197.
115. Raubenheimer, L. J., Boesman, E., and Stapleton, H. J., *Phys. Rev.*, 1965, **137**, A1449.
116. Townsend, M. G., and Orton, J. W., *J. Chem. Phys.*, 1966, **45**, 4135.
117. Harris, E. A., and Yngvesson, K. S., *Phys. Letters*, 1966, **21**, 252.
118. Harris, E. A., and Yngvesson, K. S., *J. Phys. C.*, 1968, **1**, 990 and 1011.

GLOSSARY OF SYMBOLS

The list below gives the symbols predominantly used in the text. In order to maintain the symbolism of certain published papers, it has been necessary in a few cases to use the same symbol for different quantities. In no case does this occur within a single chapter and it is thought that no confusion should occur.

α	Coefficient of thermal contact	τ_1	Phonon bottleneck relaxation parameter
β	Bohr magneton	$\chi =$	Complex magnetic sus-
γ	Gyromagnetic ratio	$\chi' - i\chi''$	ceptibility
Γ, Γ_0	Reflection coefficients of a cavity on and off magnetic resonance	χ_0	Static magnetic susceptibility
Δ, δ	Energy-level separations	χ_T, χ_S	Isothermal, adiabatic magnetic susceptibilities
Δ	Saturating pulse width (Chapter 7 only)		
$\Delta H_{\frac{1}{2}}$	Line width at half power (in gauss)	ω	Angular frequency ($= 2\pi\nu$)
$\Delta H_{pp}{}^{\circ}$	Peak-to-peak derivative line width, no saturation effects	ω_0	Resonant angular frequency
$\Delta H_{\frac{1}{2}}{}^{\circ}$	Width of the absorption line at half power, no saturation effects	ω_1	γH_1
		$a_{nm}\dagger, a_{nm}$	Annihilation and creation operators
ΔH	Residual field in rotating frame (Chapter 8)	C_M, C_H	Specific heats of spin system at constant magnetization, field
$\Delta\omega_{\frac{1}{2}}$	Line width at half power (in radians/sec)	c	Number of spins per cc
Δ_ν	Phonon bandwidth	D_0	Energy absorption by reference sample at given incident power level
θ_D	Debye temperature		
θ_{ph}	Phonon effective temperature in bottleneck conditions	D	Energy absorption by sample at same power level
Λ	Landé splitting factor	D_{ij}	Interaction tensor
μ	Spin magnetic moment	e	Input voltage to amplifier
ν	Frequency	e_{kl}	Lattice strain
ρ	Density of crystal	E	Energy
ρ, ρ_0	Faraday rotation in presence, absence of microwave radiation (Chapter 7 only)	E_{ph}	Energy of phonon system
		E_z	Zeeman energy
		f_1	Impurity concentration
		g	Splitting factor
$\rho(\delta)$	Density of phonon modes	$g(\nu), g(\omega)$	Line-shape functions
		G	Tensor whose components multiply strain to give crystal-field energy changes
σ	Bottleneck parameter		
τ	Time between pulses (Chapter 8); relaxation parameter in Casimir-du Pré equations (p. 79)		
		h	Planck's constant
		$\hbar$	$h/2\pi$
τ_c	Phonon relaxation time	H, H_0	Steady magnetic field
τ_i, τ_j	Relaxation parameters in multilevel case	H_1	Radio-frequency magnetic field
τ_{R1}, τ_{R2}	Radiative lifetimes	H_e, H_i	Effective or resultant magnetic field

$\mathcal{H}, \mathcal{H}_0,$ $\mathcal{H}_1, \mathcal{H}_{spin}$	Hamiltonian operators	S	Entropy (Chapter 5); saturation factor $= (1 + S')^{-1}$ (Chapter 6)
$\mathbf{J}$	Total angular momentum vector	S_{ij}	Saturation factor between levels i, j of a multilevel system
$\mathcal{J}$	Total angular momentum		
k	Boltzmann's constant	t	Time
$\mathbf{k}_{nm}$	Propagation vector	t_ω	Pulse duration in spin-echo experiment
$\mathbf{L}$	Orbital angular momentum vector	T	Absolute temperature
m	Magnetic dipole moment	T_1	Spin-lattice relaxation time
M	Mass (of crystal)	$T_1(\mathrm{d})$	'Direct' relaxation time
M, M_0	Magnetic moment per unit volume, magnetization	$T_1(\mathrm{O})$	Orbach relaxation time
		$T_1(\mathrm{R})$	Raman relaxation time
M_ν	Magnetization of sample under influence of saturating microwave radiation	T_2	Spin–spin relaxation time
		$T_2{}^*$	Inverse bandwidth ($= T_2$ for a homogeneous line)
$n =$ $n_i - n_j$	Population difference, not in thermal equilibrium	T_{21}	Cross-relaxation time
		T_b	Bath temperature
n_i	Number of spins in level i, not in thermal equilibrium	T_{ph}	Phonon relaxation time
		T_s	General cross-relaxation time (Chapter 7 only); Effective spin temperature
$n_0 =$ $N_i - N_j$	Thermal equilibrium population difference	T_{1e}	Measured relaxation parameter
N	Total number of spins	v	Velocity of sound
N_i	Number of spins in level i in thermal equilibrium	V	Volume of sample
		$V = \sum_{nm} V_n{}^m$	Crystal-field potential (Chapter 2)
p	Phonon excitation or occupation number not in thermal equilibrium	V_α, V_β	Potential energy
		V_c	Effective volume of resonant cavity
P	Phonon excitation or occupation number in thermal equilibrium	w_{ij}	Transition probability of cross relaxation
P_{nm}	Occupation number of phonon mode (nm)	W_{CR}	Probability per unit time of cross relaxation involving $(a + b)$ spins
P_i	Power incident on cavity	$W_{i \rightarrow j}$	Probability per spin per second of an $i \rightarrow j$ transition
P_m	Power absorbed in sample		
Q	Quality factor	W_S	Spin transition probability (stimulated)
Q_f	Normal displacement		
Q_m	Magnetic Q, due to losses in sample	$y_m{}'$	Peak-to-peak derivative curve height
r_0	Ionic radius	y, y_m	Peak absorption curve height
R	Average ionic separation	$Y_n{}^m$	Spherical harmonic of degree n and azimuthal quantum number m
S	Spin angular momentum quantum number;		

INDEX